THE NILE, BIOLOGY OF AN ANCIENT RIVER

MONOGRAPHIAE BIOLOGICAE

Editor

J. ILLIES

Schlitz

VOLUME 29

DR. W. JUNK B.V., PUBLISHERS, THE HAGUE 1976

THE NILE,
BIOLOGY OF AN ANCIENT RIVER

Edited by

JULIAN RZÓSKA

'Aut Nilus, aut nihil'
(Anon.)

DR. W. JUNK B.V., PUBLISHERS, THE HAGUE 1976

ISBN 90 6193 081 2
© Dr. W. Junk b.v., Publishers, The Hague, 1976
Cover design Max Velthuijs, The Hague
Printed in The Netherlands
Zuid-Nederlandsche Drukkerij N.V., 's-Hertogenbosch

CONTENTS

AUTHORS' ADDRESSES

BERRY, L., Professor of Geography, Clark University, Worcester, Massachusetts 01610, U.S.A.

EL HADIDI, N. M., Dr., Botany Department, Faculty of Science, University of Cairo, Giza Egypt.

ENTZ, B., Dr., Biological Station, Hungarian Academy of Sciences, Tihany, Hungary.

FAWZI HUSSEIN, M., Professor, Faculty of Science, University of Cairo, Giza, Egypt.

GHABBOUR, S., Dr., Institute of African Research and Studies, University of Cairo, Egypt.

GHALLAB, M. S., Professor, Institute of African Research and Studies, University of Cairo, Giza, Egypt.

GREENWOOD, P. H., D.Sc., British Museum, Natural History, Cromwell Road, London, S.W. 7, United Kingdom.

Hammerton, D. M.Sc., Acting Director, Clyde River Authority, Glasgow, United Kingdom.

LATIF, ABOUL-FOUTUH, Dr., Institute of Oceanography and Fisheries, 101 Kasr el Aini Street, Cairo, Egyet.

LIVINGSTONE, D. A., Professor of Zoology, Duke University, Durham, North Carolina 27706, U.S.A.

MORRIS, P., Dr., Zoology Department, Royal Holloway College, University of London, Egham, Surrey, United Kingdom.

TÄCKHOLM, V., Professor, The Herbarium, Botany Department, Faculty of Science, University of Cairo, Giza, Egypt.

TALLING, J. F., D.Sc., Freshwater Biological Association, Ferry House, Amble side, Cumbria, United Kingdom.

THOMPSON, K., M.Sc., Department of Biological Sciences, University of Waikato, Hamilton, New Zealand.

WASSIF, Kamal, Professor of Zoology, Ain Shams University, Cairo, Egypt.

WILLIAMS, TR. R., M.Sc. Zoology Dept., University of Liverpool, U.K.

WRIGHT, C., D.Sc., British Museum, Natural History, Cromwell Road, London, S.W. 7, United Kingdom.

AUTHORS' ADDRESSES

CHAPTERS' CONTENTS

The Nile in the Desert. The view encompasses more than 2000 kilometers from the river junction in the lower part of the picture to just below the Qena-Luxor bend, at the left upper end. The photograph was taken before the Aswan High Dam basin began to rise. The Red Sea and Saudi Arabia and the Ethiopian Highlands form the background. For the present state of the Dam basin see fig. 72. Courtesy of NASA.

INTRODUCTION

This book is an attempt to bring together information on the biology of the Nile. A big library could be filled by books on this river devoted to exploration and discovery, human history and hydrology. None has so far been written on the biology of the whole river system. The hazards are great, the scientific literature is scattered, most of the older papers are outdated and the new documentation covers only fragments of the river or very specific issues. A broad treatment of biological phenomena had to be adopted and only a selection of sources consulted so as to bring out the main points without encumbering the book with details.

The initiative to conceive a book on the biology of the Nile arose from the Editor's work on the hydrobiology of the Nile with a team of devoted colleagues at the University of Khartoum. On reflection, the concept of such a venture broadened for several reasons. First, the two component rivers and their tributaries flow through at least three climatic zones, four or five different landscapes and two major biogeographic regions. This is reflected in the flora and fauna of different biomes surrounding the rivers. Secondly, the whole river system is divided into geomorphologically different sectors of different origin, offering varied ecological conditions. Thirdly and most important, the Nile has for many thousands of years acted as a life artery bringing water from far away sources to now arid lands in the north. Whole landscapes have altered drastically in the last 30,000 years. The desert has finally encroached and man had to concentrate near the life artery from the late Palaeolithic onwards to the blooming fluviatile civilizations springing up in historic sequence. These are also biological phenomena of high significance and underline the importance of this unique river.

It is man who has increasingly reversed the role of the river; instead of being dominated by this life artery, he has now mastered the river to such degree that this book is already an epitaph to the Nile's existence as a free flowing water course. The prophecy of a politician (W. S. Churchill) in the early years of this century that the Nile will cease to flow into the Mediterranean has now been almost fulfilled.

No apologies are made for the shortcomings of this book nor the often controversial opinions contained. It is an attempt which had to be written now.

The book is timely. It looks back into the past and forwards to the future which has to be watched carefully as to the consequences of man's interference. This interference was necessary but is fraught with problems.

These problems have to be recognised and investigated. Though some expedition-type of work may still be necessary in some areas, it is the permanent hydrobiological centres along the Nile which will have to undertake further work in close co-operation with each other. Some of the centres are: The East African Freshwater and Fisheries Research Organisation (EAFFRO) at Jinja, in co-operation with the Universities of Makerere and Nairobi; in Ethiopia, the University of Addis Ababa; in the Sudan, the Hydrobiological Research Unit; in Egypt, the Hydrobiological Station at Aswan, the Hydrobiological Station at Alexandria and interested University departments at Cairo; all of these have to keep a close co-operation with Government departments dealing with fisheries and hydrology.

The book has contributions from a number of co-authors; all have worked at various parts of the Nile; their independent chapters had sometimes to be adjusted to the general trend of the book, and linked and cross-referenced by the Editor. He had to contribute a number of summaries of other people's work; he is responsible for any errors or misrepresentation.

Acknowledgements

Besides the co-Authors, I am greatly indebted to a number of persons:
For advice and references: F. Wendorf, U.S.A.; R. Hill, Oxford; P.
Corbet, New Zealand; A. Nibbi, Oxford; J. L. Cloudsley-Thompson,
London; Th. George, Khartoum.
For unpublished material: D. Hammerton, G. E. Wickens, Kew Gardens;
Asim el Moghrabi, O. El Badri Ali, M. Obeid, Faiz Faris Bebawi,
Khartoum.
For critical reading of chapters: J. F. Talling, D. J. Lewis, London;
G. E. Wickens, M. A. J. Williams (Australia).
For help with reference books: G. Bridson, Librarian, Linnean Society,
London.
Prof. M. Kassas has helped greatly during my fact finding visit to Egypt
in 1974.
My ardent thanks also go to Miss J. Buchanan for most generous help in
producing this book.
Permission has been granted to reproduce the following illustrations:
Figs 5, 12, 20, 23, 31, 33, 34, 59, 85, 88, 92, 93, 95, 99, 100, 101.
This is gratefully acknowledged to: The Academic Press, Blackwell
Scientif. Publications, Internationale Revue d. ges. Hydrobiologie, Lim-
nology and Oceanography, Linnean Society, Royal Metereological
Society, Royal Society, Hydrobiologia.

J. Rzóska, Editor.
6 Blakesley Avenue,
London W. 5.

I. ORIGIN AND HISTORY OF THE NILE

by

J. RZÓSKA

Three chapters deal with the important subject how the unique Nile-system arose. Four major and different 'provinces' of this enormous water course are treated in sequence: Egypt, Nubia, Sudan – Ethiopia, and the head waters. Different methods have been applied in the three contributions: geology, geomorphology, study of sediments and their contents and archaeology. Controversies exist, yet the general impression begins to emerge of a river system which is both 'ancient' in different sectors and comparatively 'recent' in their linkage.

Such an approximate picture could only be constructed through the urgency prevailing in the last 20 years in Egypt and Nubia. There the impending radical change brought about by the construction of the Aswan High Dam and its drowning of a great part of the Nile valley, resulted in a concentrated research activity with remarkable discoveries. The leader of one of the archaeological salvage expeditions, Prof. F. Wendorf, has kindly supplied material, unpublished at the time of writing. Based largely on this basis the editor has written chapter 1; errors and misrepresentations are his.

1a. THE GEOLOGICAL EVOLUTION OF THE RIVER NILE IN EGYPT

EDITOR. Under this title R. Said, of the Geological Survey of Egypt, has summed up what is known at present about the development of the Nile system in Egypt (in press, 1975). Previously, in his Geology of Egypt (1962) he has brought together the results of both the work by others and his own. From these sources an overall and very broad sequence of events has been compiled.

Sediments form one of the chief sources of information; they show great changes in the river valley since its 'down cutting' through earlier rocks in the Upper Miocene. Five river phases succeeded each other: the 'Eonile' in the Upper Miocene, 'Paleonile' in Upper Pliocene, and the 'Proto- Pre- and Neo-Nile' in the Pleistocene. These phases were separated by 'episodes', with the river declining or even ceasing its flow for reasons of tectonic and climatic changes. The ultimate origin of the Nile valley is difficult to ascertain completely because the original beds are buried under deep sediments and disturbed by tectonic 'blocks' and faults, crossing or lying parallel to the present valley. More geophysical and geodetic work would be required; some clues have been provided by numerous bore-holes all over Egypt and the offshore areas, in search for oil and water.

Middle Miocene

The northern part of the Delta was a sea bay of the Mediterranean surrounded by mountains west and south; a South Delta block was caused by faulting in late Eocene and Pre-Miocene.

Upper Miocene

There was a regression of the sea and a drying up of the northern Delta. The Mediterranean shrank into a series of salt lakes and finally dried up completely about 5.5 mill. years ago (Hsü et al. 1973). With a low sea bed level erosion was intense and the ancient Nile valley began to form, with the Egyptian Eonile cutting a deep gorge 183 m below present sea level at Aswan and 509 m at Cairo region and still deeper in the Delta. The course of this river was determined by faulting with shifts, from a north-western course to the Qatara depression, to a more northerly course.

Lower Pliocene

The sea was reconstituted by Atlantic inflows at Gibraltar and advanced along the Eonile valley as far as Aswan, drowning the valley. Sediments of this lower Pliocene contain planktonic Foraminifera in the north and

2

brackish water organisms in the south. Towards the end of this phase the sea regressed, with indications of a cool period by the presence of carbonate rich sediments. A warm climate followed about 4.5 mill. years BP with lesser carbonate contents, with subsequent phases of coolness again.

Upper Pliocene

This phase started about 3.32 mill. years BP with a deterioration of climate. The river now in its 'Paleonile' stage advanced along its valley 'cascading' over the southern Delta block into a bay of the sea; brackish organisms appear there. The Paleonile has contributed more sediments than the other stages, its water sources came from vegetation covered regions in the south e.g. from the now dry Wadis Milik, Howar and other sources from the Sudan. Some were derived from the Wadi Amur and Gabgaba in the eastern hills of the Sudan; both sources supplied an existing Nubian Nile. The Red Sea hills further north-east also contributed via the Wadi Kharit-Garara. The Plio-Pleistocene boundary of time is thought to be about 1.85 mill. years BP.

Pleistocene

In the advent of the Pleistocene great tectonic and climatic events took place. This was an epoch of high seismicity in Egypt and the Red Sea area. The previous connections with southern water supplies were severed, the western part of the Nile valley was lowered and this was the path of the future Neonile which broke through to Egypt much later. The margin of the African continent was probably in mid-Delta. The climate was arid with fluctuations, throughout the Pleistocene in contrast to the mostly wet Pliocene though there were rains in winter. Sediments show several units with intervals associated in Egypt with glacial world events. The Paleo-Proto-Nile interval from about 1.85 mill. years BP to about 700,000 was a time of great cooling connected with glaciation onset elsewhere and in Egypt also a period of dryness; an apparent stoppage of river flow into Egypt occured. The early part of this time was still pluvial with coarse sediments, terminating with the deposition of travertines in the middle of the Egyptian Nile valley. Wind deflation started and formed the large depressions in the western desert area and the beginnings of the present landscape of Egypt. This pediplenation continues until now and has destroyed almost all old surfaces; relicts remain in the Faiyum, where Paleonile sediments are seen on elevations providing a post-Upper Pliocene date.

The Protonile started probably with the onset of the glacial era. An only date in Egypt is the 320 ft. terrace described by Sandford & Arkell (1929) and dated at about 600,000 years BP. The Protonile had a more western course than the present river with sediments of gravel and sand in terraces parallel to the modern Nile valley. Water sources of this Nile were mainly from outside Egypt, possibly the Khartoum region, but the Red Sea hills also contributed via the Wadi Alaqi, Kharit and Gabgaba. This phase was apparently of relatively short duration from 700,000 to about 500,000 BP.

Prenile

No sediments have been discovered so far from the interval between Proto- and Pre-Nile. Tectonic movements and aridity severed the previous connections; new ones were formed with Ethiopia especially with the Atbara; sediments of gravel and sands differ mineralogically from previous ones with the influx of 'new' minerals (Shukri 1950). These differ also from present sediments by a higher amount of 'pyroxenes', due probably to the predominance of the Atbara over the Blue Nile at that time. The only date available is from Faiyum where a 44 and 50 m terrace has been assessed at 130,000 BP (Sandford & Arkell 1929). The Prenile was a strong river with large floods from 600,000 to about 125,000 BP, contemporaneous with the Riss glaciation. A large delta was created with sediments extending into the sea. The sediments were estimated as twice as big as those present; Ball (1939) gives the present sediment load as 110 million tons per year. The path of this river was west of the present course but east of the Protonile. The climate was arid with sand dunes inter-fingering with west bank sediments, the wadis contributed little. Deflation reached its peak at the end of the Prenile phase; the present landscape of Egypt was established. Several aggradation periods are recognised and during the first the Faiyum formed a lake 44 m above the present level. Relicts of previous sediments are now on tops of hills. Prenile sediments are known from Nubia; uplifts reduced flow of the Prenile into Egypt to a very low level about 125,000 BP.

The Pre-Neonile interval lasted from the above date to about 30,000 BP when the Neonile began to flow. In the early part of the interval, corresponding with the Würm II, there were considerable rainfalls with gravel deposits resulting. At Abbassia, north-east of Cairo, the first prehistoric remains of man of late Acheulean age were found in 1920. Some gravels had also Red Sea hill origin. This pluvial was followed by arid conditions contemporaneous with Würm I. Prenile sediments were cut down and the final shape of the present Nile valley was formed.

Neonile

This last phase started with the river 'breaking' into Egypt about 30,000 BP. This is the best studied phase with considerable documentation from radiocarbon dates; tectonic events in Ethiopia occured still and the climate was arid with some fluctuations.

4

1b. PLEISTOCENE HISTORY OF THE NILE IN NUBIA

Archaeological Evidence

Following the broad outlines of the Geological evolution of the Nile in Egypt the sequence of events in a more southern sector, in Nubia, has been revealed further through the prehistoric excavations from 1961 onwards. To those previously known, Nubia further yielded many sites of palaeolithic 'industries', dated by radiocarbon determinations and correlated with geomorphological observations and fluviatile sediments over a span of about 20,000 years pushing back the human history of the Nile valley of this region in the Sudan and Egypt. The publications by Wendorf (ed. 1968) and Wendorf & Schild (1975) incorporate most of the previously existing data.

Here only selected observations are quoted which contribute directly to the history of the Nile and its palaeoecology. The main events are brought together in a graph of changing river levels (Shiner in Wendorf 1968) and a table compiled from de Heinzelin & Wendorf (in Wendorf ed. 1968). The detailed analysis of palaeolithic 'industries' has been omitted. Within the time span from about 20,000 to 3,000 B.C. the river rose and fell as seen in the graph (Fig. 1). The main phases are named after local sites (Fig. 2) where they are best discernible.

Dibeira-Jer formation. 20,750 B.C. ± 280	Great rise of river; Fluvial deposits are 'First true Nile' (present) sediments, found 34 m above modern flood plain, 157 m a.s.l.; traces of drowned vegetation. Khormusan industry, Upper Palaeolithic.
Ballana recession 17,200 B.C. ± 275 16,650 ± 250	Water level at 133 m a.s.l.; some signs of local drought, wind erosion and dune formation. Sebilian industry and others.
Sahaba formation 12,550 B.C. ± 490 charcoal	Rise to 20 m above present flood plain, 147 m a.s.l.; numerous sites of Gemaian, Halfan, Qadan industries; richest site of aquatic molluscs; charcoal date.
Recession about 9,000 B.C.	Level drops to 129 m a.s.l.
Arkin rise 7,440 B.C. ± 180 5,750 ± 120	Rise to 135 m a.s.l.; 10 m above modern flood plain; Arkin 'industry'.

Then follows a gradual fall of river level, with fluctuations to the
Qadrus formation   Neolithic
3,650 B.C. ± 200

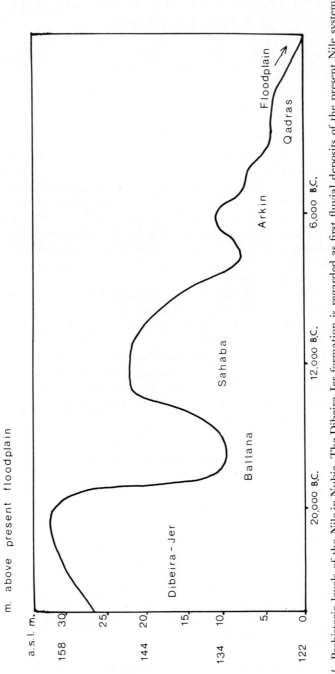

m. above present floodplain

Fig. 1. Prehistoric levels of the Nile in Nubia. The Dibeira-Jer formation is regarded as first fluvial deposits of the present Nile system; from J. Shiner, In: Wendorf 1968 vol. II.

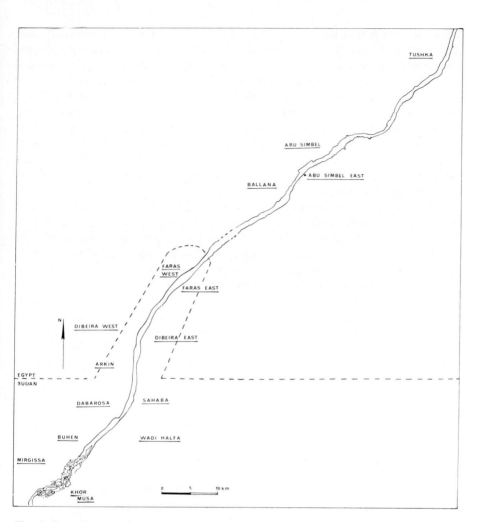

Fig. 2. Location of principal sites in Sudanese Nubia; from de Heinzelin (Atlas), in: Wendorf 1968 (changed).

Early Middle Kingdom burials and those of Meroitic and Christian times have been recognised. The fluctuations in river levels reflect palaeoclimatic events and changing river flows in the upper basin of the Nile (Atbara, Blue Nile, White Nile). The high levels may have some parallelism with glaciation phenomena in the northern hemisphere.

Sediments

Shukri (1950) has recognised the importance of the mineralogical composition of sediments for the history of the Nile system. There are differ-

7

ences between the Blue Nile and Atbara components on one side and those of the White Nile system on the other (see ch. 9). Augites of volcanic origin are conspicuous in the Atbara sediments and the Blue Nile, whereas the southern tributaries and their collecting channel the White Nile have mainly metamorphic elements in their load, with sillimanite forming 50% of it e.g. in the Ghazal river and hornblende in the Sobat representing heavy minerals. The sediment load of this system is anyway negligible. During the Pliocene and Plio/Pleistocene sediments in Egypt were devoid of augites and elements characteristic of the Blue Nile-Atbara system derived from Ethiopia; these become conspicuous by the time of the Middle Palaeolithic.

The presence and stratigraphy of palaeolithic deposits in Egypt and Nubia is still very imperfectly known but a rough sequence has been proposed by R. Said et al. (1970). The oldest at present recognisable are the Dandara silts near Luxor and Qena in Egypt with a single radiocarbon date 'greater than 39,000 BP'. (Butzer & Hansen 1968) have found marl formations near Kom Ombo with a date of 25,250 B.C. $\pm$ 1000). Extensive deposits of Ethiopian origin lie below Acheulean artifacts indicating the existence of river flow from these sources before that archaeological phase and before the present Nile regime. Gravel and sands of Eastern desert origin overlie the Dandara silts, indicating heavy rainfall in the wadis, then still active, now dry. A layer of pebbles in a red soil matrix follows, presumably again of wadi origin. A long period of erosion is discernible, forming the present topography of the Nile valley in Upper Egypt. Then follows deposition of 'slope-wash' over the landscape with numerous traces of middle and late Palaeolithic artifacts. Older sediments discovered in Upper Egypt probably represent 'small segments from a considerable time interval from early Upper Pleistocene to the middle' of that period with aggradations and recessions, but 'the sequence of these earlier events must await further work both in Sudan and Egypt' (Wendorf & Schild in press, 1975). In Nubia deposits of older sediments of Nilotic origin have not been found so far; 'non-Nilotic' or 'Pre-Nilotic' deposits from local sources are there with Acheulean material in lower strata and Middle Palaeolithic in the upper.

Even taking into account the Kom Ombo discovery by Butzer & Hansen, mentioned above, this 'did not alter the basic concept held by most of those working in Nubia that true (present) Nile sediments were deposited no more than 30,000 years ago'. The discovery of earlier Nile deposits in Egypt[1] has suggested that some of the Nile silts observed in Nubia were older than first believed; 'with the sites now buried under the waters of Lake Nasser a satisfactory answer seems impossible but it seems highly likely that several of the problem sites ... could well represent earlier episodes of Nilotic deposition' (Wendorf & Schild in press, 1975).

In two papers Fairbridge, who was a member of the Columbia

8

University Expedition to Sudanese Nubia 1961–62, has given radio-carbon dates based on mollusc shells and has made deductions on the sequence of Nile phases (1962, 1963). His opinions on river sedimentation differ from those expressed later by Wendorf *et al.* (1968). Fairbridge maintains, and Berry (ch. 2 of this book) supports him, that fall out of the sediment load of the river occurred during low water and during arid phases with diminishing discharges. Wendorf seems to connect the drop out (aggradation) with flood conditions.

Limnologists who have worked on rivers, at least in the Sudan, have observed that fall-out of suspended matter occurs whenever current velocity and with it the carrying capacity of flowing water is reduced, regardless of flood or low water conditions. This may be caused by dams or any natural obstacle; we will refer to this phenomenon observed in the Sudan and Egypt in chapter 9, and also 24 and 27.

Otherwise Fairbridge's conclusions on climatic sequences and his dates of radio-carbon tests do not alter the broad outlines of events within the scope of this book.

REFERENCES

Ball, J. 1939. Contributions to the Geography of Egypt. Egypt. Survey Dept. Cairo 300 pp.

Butzer, K. W. & Hansen, C. L. 1968 (1969). Desert and River in Nubia. Univ. Wisconsin Press.

Fairbridge, Rh. W. 1962. New radio-carbon dates of Nile sediments. Nature 196: 108–110.

Fairbridge, Rh. W. 1963. Nile sedimentation above Wadi Halfa during the last 20,000 years. Kush 11: 96–107.

Hsü, K. H., Ryan, W. B. F. & Cita, M. B. 1973. Late Miocene desiccation of the Mediterranean. Nature 242: 240–244.

Said, R. 1962. The Geology of Egypt. Elsevier, Amsterdam, New York.

Said, R. 1975. The Geological Evolution of the river Nile in Egypt. Proc. Intern. Conf. North East African and Levantine Pleistocene prehistory. S. Methodist Univ. Dallas, Texas 5–8 Dec. 1973. Preprint of 1975 publication.

Said, R., Wendorf, F. & Schild, R. 1970. The Geology and Prehistory of the Nile Valley in Upper Egypt. Archaeologia Polona 12: 43–60.

Sandford, K. S. G. & Arkell, W. J. 1929. Paleolithic Man and the Nile Faiyum divide. Chicago Univ. Oriental Inst., Public. 1: 1–77.

Shukri, N. M. 1950. Mineralogy of some Nile sediments. Quart. J. Geol. Soc. London 105: 511–534.

Wendorf, F. ed. 1968. The prehistory of Nubia. Vol. I, II. Fort Burgwin Res. Centre and South. Method. Univ. Press.

Wendorf, F. & Schild, R. 1975. The Paleolithic of the Lower Nile Valley. Preprint as R. Said 1975.

[1] See Epilogue.

9

2. THE NILE IN THE SUDAN, GEOMORPHOLOGICAL HISTORY[1]

by

L. BERRY

The landmass of Africa is old and the river systems have evolved over long periods of time. Most of the great rivers occupy basins which in part have been carved out by the rivers but are also large down-warps in the earth's crust. The broad character of river basins have thus evolved over tens of millions of years of geological time and involved major geophysical events; while the details of the river's course have often evolved within a few tens or hundreds of thousands of years of time and mostly are recorded in erosional-depositional events.

The Nile is no exception to this general rule, though there are many unique features of its basin. The Nile in its present form appears to be made up of several distinct systems (Ed.: already recognised by Lyons 1909)[2] which become joined at a late stage in geological history (see also ch. 1). Each part of the system is related to a different structural and historical setting. The Blue Nile basin has been formed and modified in relation to the uplifting of the Ethiopian plateau, which was accompanied by massive volcanic outpourings. The headwaters of the White Nile drain the tectonically and volcanically active Ugandan plateau area where many detailed drainage changes have been caused by rift-associated geophysical activity. The White Nile basin in the Sudan occupies a very broad and, in places, ill-defined tectonic depression including the in-filled Sudd region. Lastly, the main Nile north of Khartoum, though less obviously structuredly controlled, flows north in the tectonic low behind the uplifted rim of the continent – the Red Sea Hills.

The interplay of long continued tectonic activity, river processes of erosion and deposition, and the effects of climate fluctuations, certainly over the last 1,000,000 years, combine to produce a complex history. Although some of that history is becoming known, much is still missing. The history of the Nile in the Sudan is still largely locked in the sediments and erosional geomorphology of the basin. It is most conveniently discussed by considering the Blue Nile, the White Nile and the Main Nile as separate systems.

---

[1] Parts of this chapter are based on an earlier paper 'The Nile in the Sudan', Berry, L. & Whiteman, A. J., Geog. Journal, 1968.
[2] Ed. see fig. 4 and quotation at the end of ch. 2.

The geological and geomorphological evidence clearly indicates that the Blue Nile is an ancient river system, contrary to the views expressed by Arkell (1949) and others. Rivers must have existed on the flanks of the Sudan section of the Afro-Arabian Swell and the roughly parallel courses of the Blue Nile, Dinder, Rahad and Atbara probably point to some overall tectonic control, perhaps dating back to the interval Late Cretaceous-Late Eocene when the swell was formed (Whiteman 1971). Vast quantities of lava were extruded during the Oligocene in Ethiopia and the eastern Sudan, and the river system as we know it today probably originated on this vast volcanic pile. Of Miocene and Pliocene denudation chronology we know very little. Lake Tana is supposed to have been formed in the Pliocene (Dainelli 1943; Mohr 1962) and the Abbai canyon was excavated during the Pliocene and Pleistocene, forming the broad sweep around the volcanic centre of Mt. Birhan[1].

The formation of the Ethiopian plateau must have caused changes in the atmospheric circulation, probably accompanied by increased precipitation and run-off. Whether at all times this was sufficient to provide a permanent, integrated, river system, or whether at times each river ended in a delta, as the Gash or Mareb does today, is difficult to determine. A reduction of the amount of rain falling in the Blue Nile or Atbara catchments could result in the rivers failing to reach the White Nile or the main Nile, and this may well have happened to the Atbara during Pleistocene dry periods. The thick superficial deposits of the Atbara near the Butana Bridge below Khasm el Girba and the broad alluvial fill of the Atbara between Qoz Regeb and Atbara town are probably aggradation deposits formed during dry periods when the river was incapable of flushing its suspended load down to the main Nile.

The Blue Nile in Sudan flows across its own sediments from Roseires northward. Narrow to the south, the valley widens northwards and from about the latitude of Sennar merges with the Gezira Plain, though the river is still incised well below the general topographic level.

The Gezira, with predominantly clays at the surface, slopes very gently towards the north and west and possesses many characteristics of an inland delta graded to the White Nile. Sandspreads and dunes occur in places and aerial photographs and field studies show that these patches are associated with an old distributary system which was probably formed by overbank floods during Recent and Pleistocene times (Williams & Adamson 1973).

A most striking feature of the Gezira clay plain is the comparative uniformity of the thick 'Gezira Clay', which is broken only by occasional sandy patches and sand dunes. Photo-interpretation and fieldwork

---

[1] See new evidence in Epilogue.

reveal that these patches are associated with shallow discontinuous channel systems. It seems clear that overbank flooding through these channels from the Blue Nile towards the White Nile was an important feature of the palaeohydrology of this area, and that the sandy areas are remnants of material laid down in channels.

Williams & Adamson (1973, Fig. 1) show from a study of fossil shells and heavy minerals that the Gezira materials were deposited by 'heavily laden fast flowing seasonal rivers … prolonged aggradation by seasonal rivers carrying coarse loads would fill pre-existing depressions and would eventually result in the formation of a complex and heterogenous alluvial fan.'

At present most of the flood water is carried through the region in the incised channel, although some does spread into backwaters and abandoned meander channels. As is well known, the Blue Nile has a marked seasonal flow. During low water little material is carried in the water but during the flood there is a large suspended load (up to 3,000 parts per million) of clay and fine sand, mostly derived from Ethiopian soils and sediments plus a significant dissolved load (see ch. 9). Also during the flood considerable reworking of sandy and pebbly bedload occurs and sand bars and islands are in continual evolution. As the flood recedes some of the suspended load is dropped on the midstream islands and meander curves, and on the higher part of the bed as it is exposed during falling water. This clay and silt forms a fertile strip within the main channel.

The White Nile

The White Nile basin in Uganda was possibly only joined northward as late as the Pleistocene period (see ch. 3). The comments here refer to the river north of Juba in Sudan.

North of Juba the Bahr el Gebel flows over the outcrop of the Umm Ruwaba formation (? Tertiary and Pleistocene) which consists of fluviatile and lacustrine sediments laid down in a series of land deltas with extremely low surface gradients, much as the Gash Delta, north-west of Kassala, is being formed today. Details of thickness, age, etc. of this formation are uncertain. Where the Nile debouches onto the Sudan plains north of Juba it flows in a defined valley as far north as Mongalla The valley walls decrease in height northward and the series of sections shows the flood plain gradually merging with the surrounding country until north of Mongalla the banks slope gradually away from the river. North of Juba there are several channels each with marked levees. Numerous out-flow points occur along the levees and, during floods, water spills into the depressions between, and forms many temporary lakes. The levees are mainly formed of fine sand whereas the back-swamp deposits are composed of silt and clay with occasional sandy lenses. Further north, as the proportion of sandy material carried by the river decreases, the levees are smaller and are less significant. Between Mongalla and Lake No overbank floods occur frequently during high water and water accumulates in swamps and small lakes. Sediments in these areas are clays and fine sands. (For the biological effects see ch. 12).

This is the typical Sudd Region, though it should be emphasized that there is a marked downstream slope, and that in the cleared river channels flow is quite strong, as much as 1.1–1.3 m/sec. in the upper main stream. The swamps are therefore caused by overbank floods and not by longitudinal ponding of the rivers as is commonly thought. At present the flow of water through the Sudd is controlled largely by the height of Lake Victoria. Increased rainfall and discharge from the highlands of the southern Sudan during the Pleistocene must have had an effect on the development of the Sudd formation, as did desiccation during the Late Würm (Fairbridge's Siltation Phase, Fairbridge, 1963).

The Bahr el Arab joins the Bahr el Ghazal at the northern end of the swamps, but unfortunately we know almost nothing about the geomorphology and geology of this system.

South of Malakal the Bahr el Jebel is joined by the Sobat which rises in the Ethiopian Highlands. The Sobat has a strong seasonal flow similar to that of the Blue Nile. Little is known about its basin or its history except that it must have been similar to that of the Blue Nile and may have originated in post-Oligocene times on the surface of the Ethiopian Traps.

The White Nile from the Sobat junction to Khartoum is a most unusual river with a gradient far lower than that of the Sudd, the fall being only 38 ft. (11.6 m) in 500 miles (800 km) (1 in 70,000) with a corresponding decrease of current. For most of this distance it flows in a well defined channel or channels within a broad valley, which suggests that this is an old and fundamental drainage line. In much of this section the river is braided and despite the low gradient there are few meander curves. The various channels divide and rejoin, and there are many low alluvial islands. Berry (1962a, b) suggested that the smaller islands are forming at present and the larger islands were formed in the past when discharge was higher. This has since been substantiated by a radiocarbon date of 3,070 BP (Williams & Adamson 1973) and of 5500 ± 90 (Adamson et al. 1974) for shells on some large islands.

The White Nile, until it reaches Khartoum, flows at a lower level than the Blue Nile and there is a pronounced slope between them. Terraces occur on both banks and two main levels have been recognized at 386 m. (1,235 ft.) and 382 m. (1,222 ft.) a.s.l. The terraces appear to be horizontal and probably mark the shores of long narrow lakes which occupied the lower valley. The 386 m. (1,235 ft.) terrace has been traced from Omdurman as far south as the Melut bend. It is distinguished both by its topographic form and by its distinctive soil characteristics (Berry 1962c; Grove & Goudie 1971). The lower terrace occurs between Khartoum and Renk. Beach-like accumulations of gastropod shells, first noted by Tothill (1946), occur in places near the backs of these terraces. Arcuate sand ridges, which were probably formed as lake shore beaches, are prominent features south of Renk and near Kosti. The lower lake appears to have been about 12 miles wide in the Gebelein area, while the higher lake may have been up to 25 miles wide, but both were narrow in relation to their lengths which must have been between 300 and 400 miles.

Main Nile

At Khartoum alluvial deposits are much more restricted and the main Nile flows northward largely over the Basement Complex and Nubian formations, except in the Shendi and Kerma basins where there are extensive tracts of alluvium. In the Khartoum area the thickness of alluvium is highly variable and in places there are infilled pools where the alluvium is some 60 ft. thick. Clay and gravel terraces also flank the river in the Khartoum area, but rapidly narrow northwards towards Geili Station.

Interpretation of the terrace deposits between Khartoum and the Sixth Cataract at Sabaloka (fig. 3) is difficult because of obscuring wash derived from the Nubian and Basement formations. The higher terrace deposits are best exposed mainly on the west bank and have been studied

Fig. 3. Part of the Sabaloka Gorge, airview. (Aero Films).

15

Table 1.

Name	Height in Meters	Age in years B.P.
Esh Shaheinab Neolithic site av.	377	5,253 ± 415
Khor Umar gastropod site	385	7,400 ± 120
Sample No. 1139, 12 km south of Abu Hibeira, 13° 56′N, 32° 33′E	377.6	8,730 ± 350
Sample No. 1211, 6 km east of Esh Shawal, 13° 35′N, 32° 41′E	379.45	11,300 ± 400

at Omdurman, Khor Umar, Esh Shaheinab, Geili and the south entrance to the Sabaloka Gorge. Acheulian and Late Acheulian, Sangoan, and Lupemban implements have been recorded from these gravels (Arkell 1949; Cole 1964).

A number of Neolithic sites have been recognized along the banks of the White and Main Niles between Gebel Aulia Dam, south of Khartoum, and the Sixth or Sabaloka Cataract. Esh Shaheinab is the most famous of these (Arkell 1949). Many vertebrates were collected and according to Bate (in Arkell 1953) the faunal association suggests somewhat drier climatic conditions than those which prevailed at early Khartoum but still wetter than those of the present day (see Palaeoecology ch. 4a). A list of radiocarbon dates for the White and Main Niles in the Khartoum area is given below, not all Neolithic (Table 1).

North of Geili the key to an understanding of many aspects of the river is the realization that the Nile is in various stages of superimposition of its bed and valley from Nubian Sandstone sedimentary rocks onto a very varied set of rocks of the Basement Complex. For example, the key to the denudation history of the Sabaloka area lies in the occurrence of Nubian Sandstone formation on the tops of Jebel Milakit. Clearly the Sabaloka inlier was covered by the Nubian formation and it seems probable that the early course of the Nile was initiated on a surface cut across the formation. The whole process has occupied vast periods of time. Down-cutting may have started in Late Cretaceous times soon after the Nubian formation had been deposited and the early Nile flowed northwards to the sea in the latitude of Aswan where marine Cenomanian rocks occur. Erosion continued in the Early Tertiary and the Shendi erosion surface was formed.

During Early Tertiary time (? Oligocene) a depression developed over the present Nile valley, north of Sabaloka, and this was occupied by a freshwater lake in which the Hudi Chert formation was laid down. With the lowering of sea level consequent on the extensive retreat of Tethys northwards in Oligocene times the early Tertiary Nile slowly cut through the Hudi Chert, the silicified sandstone and the rest of the Nubian for-

16

mation, and eventually became 'locked' in the igneous and metamorphic rocks of the Basement Complex, its detailed course being controlled mainly by joints and faults in the Basement Complex. Ultimately in Pleistocene times the river must have attained almost its present form. The river is in places flanked by wide terraces (2–3 miles wide) e.g., south of Shendi; where the Nubian escarpment, the free-face weathering back from the river, is as much as 35 miles from the bank and over 180 ft. high.

The details of the history of the Nile in Northern Sudan have been the subject of much debate. Wendorf *et al.* (1965, p. 28) gives a geological sequence for the superficial deposits of Sudanese Nubia which differs fundamentally from that given by Fairbridge (1962, 1963) and to some extent from that given by Sandford & Arkell (1933), who associated periods of silt accumulation with periods of desiccation. The Siltation Phase was initiated some 20,000–25,000 years ago. Fairbridge thought that at times flow to the north dwindled almost to nothing. Heinzelin & Paepe (1964) suggest that there is a much more complicated sequence than that suggested by Fairbridge. The Sebilian Silts were not formed under flood conditions and like Fairbridge (1963) I think that they must have formed during arid periods when the river was sluggish and incapable of transporting its suspended and traction loads[3] in Sudan and Upper Egypt. It is indeed unfortunate that such problems remain unresolved, as so much of the area in question has already disappeared under 'Lake Nasser'!

The early chronology of the Nile must necessarily remain vague, as evidence is fragmentary, but in studying the most recent part of the river's history radiocarbon dating provides some important time scales. Williams & Adamson, 1973; Grove & Warren, 1968; and Whiteman, 1971 all have summarized their conclusions on this.

Williams & Adamson show that the 382 m White Nile valley Lake was in existence 8,000–12,000 years B.P. and Grove & Warren confirm that at this time there were widespread high lake levels in Africa. West of Gebel Aulia small lakes were common 8,500–7,000 years ago (see also Williams, Medani, Talent & Dawson, Sud. Notes and Records 1974) and at that time the Nile was 6–7 m higher than at present. About 3,000 years ago the White Nile was 1–2 m above its present position. Despite the many C^{14} dates for the Nubia area of the main Nile we are still far from establishing a coherent stratigraphy for that area.

REFERENCES

Adamson, D., Clark, J. D. & Williamson, N. A. J. 1974. Barbed bone points from Central Sudan and the age of the 'Early Khartoum' tradition. Nature 249: 120–123.

---

[3] Ed.: See remarks at end of ch. 1b.

Arkell, A. J. 1949. The Old Stone Age in the Anglo-Egyptian Sudan. Sudan Antiquities Service I: 1–52.

Arkell, A. J. 1953. Shaheinab. Oxford University Press.

Berry, L. 1962a. Alluvial islands in the Nile. Rev. Geomorph. Dyn. 12.

Berry, L. 1962b. The characteristics and mode of formation of the Nile islands between Malakal and Sabaloka. Eighth Annual Rept. Hydrobiological Research Unit, University of Khartoum, 7–13.

Berry, L. 1962c. The physical history of the White Nile. Ibid.

Cole, S. 1964. The prehistory of East Africa. Weidenfeld and Nicolson.

Dainelli, G. 1943. Geologia dell' Africa Orientale. (3 vols. text and 1 vol. maps). R. Acc. Italia, Roma.

Fairbridge, R. W. 1962. New radiocarbon dates of Nile sediments. Nature 196: 108–110.

Fairbridge, R. W. 1963. Nile sedimentation above Wadi Halfa during the last 20,000 years. Kush 11: 96–107.

Grove, A. T. & Goudie, A. S. 1971. Late Quaternary lake levels in the Rift Valley of southern Ethiopia and elsewhere in tropical Africa. Nature, 234, 403–405.

Grove, A. T. & Warren, A. 1968. Quaternary Landforms and Climate on the South side of the Sahara. Geog. Journ. 134, 194–208.

Heinzelin, J. de & Paepe, R. 1964. The geological history of the Nile valley in Sudanese Nubia; preliminary results. Contribution to the prehistory of Nubia assembled by F. Wendorf.

Mohr, P. 1962. The geology of Ethiopia. University College of Addis Ababa Press.

Sandford, K. S. & Arkell, W. J. 1933. Palaeolithic man and the Nile Valley in Nubia and Upper Egypt. Orient. Inst. Publ. 17. Chicago.

Tothill, J. D. 1946. The Origin of the Sudan Gezira clay. Sudan Notes and Records 27: 153–183.

Wendorf, F., Shiner, J. L., Marks, A. E., Heinzelin de, J. & Chmielewski, W. 1965. The Combined Prehistoric Expedition: Summary of 1963–64 field season. Kush 13: 28–55.

Whiteman, A. J. 1971. The Geology of Sudan. Oxford U.P.

Williams, M. A. J. & Adamson, D. A. 1973. The Physiography of the Central Sudan. Geog. Journal Vol. 139: 498–508.

EDITOR'S ADDITION

Lyons, H. G. 1909. Longitudinal Section of the Nile. Geogr. J. 34: 36–51.

This is 'not the course of a normally developed river, where a mountain tract of rapid erosion passes gradually into a valley tract with slight erosion and into a plain where deposition is going on. The present Nile must include portions of river-systems of very different origin and date. While the drainage lines of the equatorial plateau are comparatively recent, the meeting ground of the Bahr el Gebel, Ghazal, White Nile and Sobat must have been part of a much older systems' (Lyons 1909, p. 50).

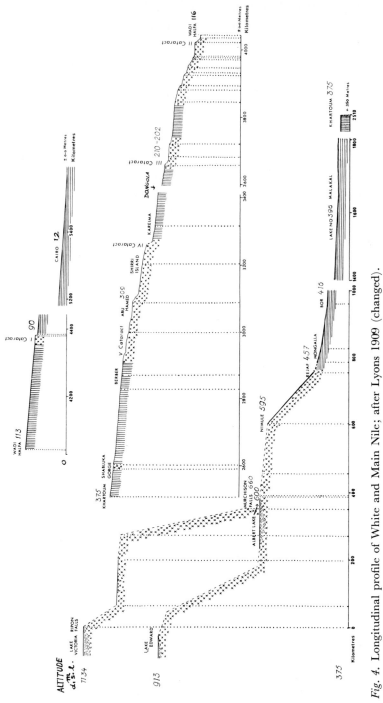

Fig. 4. Longitudinal profile of White and Main Nile; after Lyons 1909 (changed).

19

followed by a long period of crustal stability before several episodes of doming produced tilted strandlines in the Upper Pleistocene, but more likely the Victoria basin was formed as well as tilted during the latter part of Pleistocene time.

Some workers have believed the lake to be much older, and have postulated a Miocene origin. This idea is based on the outcropping of Miocene and early to middle Pleistocene beds around Kavirondo Gulf[1] at the northeast corner of Lake Victoria. These beds, however, extend to altitudes well above the rim of the Lake Victoria basin, and it is doubtful that they are related genetically to the modern lake (Bishop 1965).

Kavirondo Gulf is bounded by faults, and water wells drilled close to the northern edge of the Kavirondo basin, and somewhat to landward of the modern lake shore, have penetrated more than five hundred feet of fine-grained material characterized as lake sediment in the driller's log (Saggerson 1952). I have examined washed cuttings from one of these wells and am unable to find any aquatic microfossils, but this is not compelling evidence against a lacustrine origin for the beds. If they are lake deposits, they probably represent a considerably greater span of time than has elapsed since the Middle Pleistocene.

It is possible that Speke Gulf occupies an originally autonomous lake basin of similar great age, and that both joined Lake Victoria when it tilted east during Sangoan time.

In addition to the tilted strandlines, Lake Victoria is surrounded by three that are horizontal, at 3, 12 and 18 m above the 1960 lake level (Temple 1964). At least the upper two of these strandlines probably represent stages in downcutting of the Nile outlet at Jinja and are without palaeoclimatic significance. All three were believed independent of climatic change until a series of three wet years in the early nineteen-sixties raised the lake perilously close to the level of the lowest raised strandline. Only this lowest strandline has been carbon dated. Charcoal with an age of 3,720 years occurs in one of its beaches at Hippo Bay near Entebbe, Uganda (Stuiver et al. 1960).

The palaeolimnology of Lake Victoria was studied by Kendall (1969) whose most complete section was an 18 m core under 9 m of water from Pilkington Bay on the north shore of the lake not far from the Nile outlet. This core was carbon-dated in 28 places, which enabled Kendall to compute the rate at which microfossils had been sedimented. He was therefore able to escape the ambiguities that surround most microfossil analyses, where the abundance of each taxon can only be expressed as a percentage of the total microfossil assemblage at each level.

The core consisted of rather uniform organic lake mud down to a level with a radiocarbon age of 14,000 years, where there was an unconformity. Below the unconformity the sediment was also organic, but less so, and

---

[1] Ed. see map in ch. 11.

had a very much reduced water content. This gross stratigraphy showed that the water level 14,000 years ago was at least 26 m lower than it is today. Such a drop would bring lake level well below the outlet, cutting off the flow of the Victoria Nile.

By his studies of exchangeable cations in the mud and of minerals, primarily carbonates, deposited before and for some time after the date of the erosional unconformity, Kendall was able to show that the lake had been closed from the beginning of his record until 12,500 years ago. Although he found no indications of saline conditions at Pilkington Bay, the lake was clearly more alkaline than it is now during the long period of closure, and its phytoplankton community differed from the modern one. *Stephanodiscus astraea* was more abundant than it has been in more recent times, indicating a lower concentration of dissolved silica in the water (Kilham 1971). (For present phytoplankton see ch. 27 by Talling).

There was apparently one subsequent episode of closure, around 10,000 years ago, although it was not sufficiently long-lived to affect the chemistry of the water to a detectable degree. A second core taken in shallow water closer to the shore of Pilkington Bay lacked lake sediments older than 10,000 years, although it apparently penetrated to non-lacustrine rocks. The simplest explanation for this stratigraphic finding is a fall in lake level 10,000 years ago with removal by waves of the earlier sedimentary record.

Kendall's investigation included a very thorough analysis of the Lake Victoria pollen assemblages of the past 14,700 years. Because his radiocarbon control permitted him to use pollen deposition rates, his vegetational conclusions are the most secure of any that have been obtained in Africa.

Prior to 12,500 years ago the pollen spectra were dominated by grasses. It seems reasonable to conclude, as Kendall does, that the vegetation was some sort of savanna, although neither grassland nor woodland can be ruled out conclusively. When the lake level rose to establish the Nile outlet, rain forest tree pollen became much more abundant in the fossil record, gradually at first, and with a temporary setback around 10,000 years ago, but culminating in a period of maximum abundance from 9,500 to 6,500 years ago. About 6,500 years ago the climate became drier or more seasonal or both, and indicators of semi-deciduous forest increased at the expense of evergreen forest taxa until about 3,000 years ago. Since then all tree pollen types have gradually declined without any compensatory increase in grasses or other grassland, woodland or savanna indicators. This change is probably due to the advent of agriculture; the indigenous crops of Uganda are poor pollen producers and do not register in the pollen record.

Kendall's data show beyond any possibility of doubt that Lake Victoria was a basin of internal drainage from at least 14,700 until 12,500 years ago. A most unlikely tilting back and forth and back again would be required to explain the sedimentary record in terms of tectonic activity, so one is left with a climatic explanation. The climate must have been dry prior to 12,500 years ago, and wetter since. Such a climatic history is completely consistent with the pollen data, which are independent of the sorts of lacustrine changes that could be produced by river capture or crustal warping.

It is not possible to tell from Kendall's results how far the water level may have fallen below the bottom of Pilkington Bay. In 1972 by courtesy of the Director, Dr. John Okedi, and Capt. G. S. Illugason of the East African Freshwater Fisheries Research Organization Laboratory, I took piston cores from a number of deep-water stations in northern Lake Victoria. The deepest of these, a core 10 meters in length taken below 66 m of water, ended in coarse shelly sediment indicative of deposition under high-energy conditions. This shows that the level of the lake fell some 75 meters below that of modern times, but there is another 13 meters of water, and presumably some meters of sediment as well, in the deepest part of the lake. Until an erosional unconformity is found beneath the deepest water we cannot be sure that the lake dried up completely. The remnant, if there was one, was shallow and probably rather alkaline.

This coring establishes neither the age of the basin nor the length of time that it has held water continuously. Kendall believes that he exhausted the record at Pilkington Bay, but this is likely to be one of the younger parts of Lake Victoria, flooded for the first time by the eastward tiltings that produced the tilted strandlines. The western part of the lake probably contains older sediment than any that has yet been studied, and a long lacustrine record may even underlie the shell zone of my piston core from 66 m.

These uncertainties are particularly regrettable because of the relevance of lake age to the evolution of the cichlid species flock. The existence of five endemic species in Lake Nabugabo (Greenwood 1965), a beach pond that cannot have been separated from Lake Victoria for longer than 3,720 years, indicates that evolution can be very rapid in this family. It does not seem likely, however, that the whole flock of endemic cichlid species now assessed at 150–170 (Greenwood 1974) should have originated during the 14,000 years since Pilkington Bay was dry, and the fish fauna certainly is suggestive of the long-continued existence of a lake somewhere in the Victoria Basin. The most likely place to find evidence for such a long lacustrine history is the western part of the lake, but perhaps one should not rule out the possibility of a long continuous existence of Kavirondo Gulf as a separate lake before it was joined by Lake Victoria. Kavirondo is very shallow, and not separated from the main lake by any sill, but a sill could have been removed by the advancing waves of Victoria, and it is likely that orographic rain on Mt. Elgon would provide runoff to a Kavirondo lake through the climatic vicissitudes of the Pleistocene.

Lakes of the Albert Nile

Lake cores taken on and around the Ruwenzori Mountains shed some light on the hydrologic history of the other major tributary of the White Nile. Livingstone (1967) cored three high altitude lakes for pollen analyses. Hecky & Degens (1973) have taken piston cores from the Zairean waters of Lakes Edward (now Idi Amin Dada) and Albert (now Mobutu Sese Seko)[1]. Mr. Thomas Harvey of Duke University is currently studying the diatom stratigraphy of a suite of cores from the Uganda waters of Lake Albert. I am indebted to Dr. Hecky and Mr. Harvey for permission to summarize some of their unpublished results.

The cores of Hecky & Degens from the two large Rift Valley lakes suggest, particularly by their calcium carbonate content, an increase in the concentration of the lake water at a time between two and four thousand years ago. The Lake Edward core represents only six thousand years, and is without further present interest, but the Lake Albert core terminated in a dry, carbonate-rich layer with an age of about 13,000 years. This dry layer is situated 51 meters below modern water level, and indicates that the water was at least that much lower 13,000 years ago. Mr. Harvey has found similar, but even more convincing, indications of lowered lake level along a transect extending out from Butiaba to the 46-meter contour. An attempt by Hecky & Degens to test for evidence of drying in the deepest part of the lake was frustrated by the very rapid rate of sedimentation there. Their piston corer would not penetrate beyond sediments with an age of 7,000 years.

A detailed record of changes in Lake Albert will soon be available from the diatom analyses of Mr. Harvey. In the meantime, it is clear that the water level rose to establish the present outlet about 12,500 years ago, the time of establishment of the Victoria Nile. Mr. Harvey's oldest core covers the past 28,000 years and suggests that the period of internal drainage in the Albert basin was short-lived. The lake appears to have been open until 13,340 years ago and to have opened again by 12,480 B.P.

Lake Rudolf

Butzer and his colleagues (Butzer et al. 1969; Butzer 1971; Butzer et al. 1972) have found in the delta of the Omo River, which provides 80–90 percent of the runoff to Lake Rudolf, good evidence for a complex series of lake-level changes. The lake was 60 to 70 meters above its modern datum at a time somewhat more than 35,000 years ago, but from 35,000 until 9,500 years ago it must have been low, for neither deltaic nor littoral sediments from that age span have been found. From 9,500 to 7,500 years ago the lake fluctuated from +60 to +80 meters, and then

---

[2] In the following pages these lakes will be called by their old names for brevity (Ed.).

shrank to its present size until shortly before 6,600 B.P. when it reached +65 m, going on to reach +70 m about 6,200 years ago. This level was maintained until 4,400 B.P. when a temporary regression occurred, followed by a final transgression to +70 m a little before 3,000 B.P. Since then the level has been relatively low, but fluctuating through a range of over 40 m. It fell from +15 to —5 m between 1897 and 1955 (Butzer 1971), and rose, like most East African lakes, from 1962 to 1965.

The recurrent +70 m level strandline corresponds in altitude to a swampy overflow channel to the Lotigipi mud flats to the west. The temporary maxima around +80 m between 9,500 and 7,500 years ago probably induced overflow from the Lotigipi flats to the Pibor, Sobat and Nile rivers. Their zoogeographic affinities (see ch. 7, by Greenwood) indicate a connection between Rudolf and the Nile. It is not known how long the distinctive Nilotic fauna of the lake, such as the Nile perch, will be able to survive the present regime of evaporative concentration of Lake Rudolf water. If this time is short compared to the times that the lake has been isolated in the past, Rudolf is essentially removed as a possible site of allopatric speciation of any appreciable part of the Nilotic fauna.

The pollen analyses of Bonnefille (1970, 1972a, 1972b) show that climatic oscillations have been a feature of the Rudolf basin for more than two million years.

Other Lakes

Morrison (1961, 1968; Morrison & Hamilton 1974) cored Lake Bunyonyi and a number of former lakes now overgrown with swamp in southwestern Uganda for pollen studies. He does not seem to have found indications of changing water level in this currently well-watered country.

Outlook for future research

Several aspects of the palaeolimnology of the Nile demand attention. Longer cores from the deepest water of Albert, Edward and Victoria are needed very badly to show if these lakes persisted through the last dry period, and whether chemical conditions in them were such as to permit the survival of lacustrine fishes.

Lake Tana should provide information about the palaeohydrology of the Blue Nile similar to that which Albert provides for the White. It is not so far downstream from most of the catchment of its branch of the river, but it might give a valuable index to the palaeohydrology of the Nile's larger branch. Dr. Bonnefille has under pollen analysis a short core from the lake. The material is not rich in pollen, but would, if the core were long enough, settle the basic question of whether Tana were closed or open during the last interpluvial.

Runoff from Lake Rudolf to the Nile is indicated by delta deposits, raised strandlines and overflow channels, but the role of Rudolf in the Nile system cannot be understood without cores from the waterlogged fine sediment under deep parts of the lake. Such cores should yield information about salinity changes, and permit evaluation of Rudolf as a permanent habitat for lacustrine fishes. They would also provide environmental background for studies of human evolution and archaeology.

All of these main headwater lakes seem to be capable of providing a much longer record than they have yet. Albert and Tana in particular, seem likely to contain a record for all of Pleistocene time and more. Such a record would have great palaeoclimatic interest, but less significance for Nile palaeohydrology than the problems outlined above.

The existing lakes on the downstream reaches of the Nile are artificial reservoirs too young to be palaeolimnologically interesting, but there are three bodies of lake sediment that would repay study. Geomorphologists (Berry & Whiteman 1968) have effectively demolished the notion of a large 'Lake Sudd', but the deposits of the existing 'Sudd', the Upper Nile swamps, though presumably dating only from times since the end of the last interpluvial, deserve careful stratigraphic investigation. The lake beds extending four or five hundred miles up the White Nile from its confluence with the Blue, which Berry & Whiteman (1968) attribute to damming behind a plug of Blue Nile silt when the White Nile resumed its flow after the end of the last interpluvial, present an intriguing problem (now in progress, M. Williams in litt.) in the palaeolimnology of a dynamic short-lived lake. The Fayum depression, though only intermittently and sometimes artificially connected to the Nile (Said 1962), should present a long palaeolimnological record. Its many strandlines at altitudes between +40 m and —2 m a.s.l. do not seem to have been carbon-dated.

Conclusion

Lakes are a very characteristic feature of the Nile system. Some provide a habitat for lacustrine species and buffer the discharge against even greater extremes of seasonality. Lakes constitute a sediment trap as well, catching suspensoids and bed-load from the upper reaches of the Nile drainage. This is particularly important for the White Nile, because the main highland areas from which most of the eroded sediment comes are all located upstream from the lakes.

Before 12,500 years ago the climate was drier than it has been since, and water levels were, briefly in the case of Lake Albert and for a longer time in the case of Victoria, so low that these lakes and their catchment areas were disconnected from the White Nile. This greatly diminished the length of the Nile, and although the total effect on discharge was slight, it likely accompanied some decrease in the much more voluminous flow from Ethiopia. The effect on the river of removing the White Nile contribution was far out of proportion to the total drainage loss, because it is largely that minor contribution which sustains the flow of the lower river during the season when Blue Nile flow is minimal. Probably the total discharge was considerably reduced during the interpluvial period but the dry season discharge very much more so. This reduction would limit the size of biota that could survive from early moister times.

Between 7,500 B.P. and 9,500 B.P. there seem to have been a number of times when the level of Lake Rudolf rose to overflow via the Pibor and Sobat into the Nile. This is an adequate explanation for the nilotic cast of the modern fauna of the lake. Discharge from of other parts of the Nile drainage is likely to have been much enhanced at the same time.

Palaeolimnological methods can determine the level of the source lakes during peak pluvial conditions but not the cross-section of the outlet streams except possibly in arid-land lakes like Rudolf, where the overflow channel is unlikely to have changed shape appreciably under later erosion. This has not yet been accomplished, even for Rudolf.

It is not known, though it is in principle determinable, if the head-water lakes dried completely, or if they held sufficient fresh water to provide a habitat for such lacustrine species as cichlid fishes through each interpluvial. The biogeographic evidence might be argued both ways, with the endemic cichlids of Victoria and Edward indicating freshwater survival, the lack of many nilotic fishes, and, in the case of Edward, Nile crocodiles as well, indicating complete dryness; there is a controversy here with Greenwood (see ch. 7). In the case of Tana we do not even know if the water level fell below the outlet.

These climatic changes have had very complex geomorphological consequences in the lower Nile valley. Large areas of the Nile catchment receive rainfalls greater or less than that which results in the highest rate of erosion (Langbein & Schumm 1958), so an increase in rainfall over the entire basin would increase erosion in some areas, decrease it in others.

Tributaries draining lakes would carry decanted water, while other tributaries would carry a full complement of bed-load and suspended sediment and climatic changes need not have been synchronous over the entire Nile basin. It is hardly surprising that there have been differences of opinion about the climatic interpretation of aggradational and de-gradational features in the Nile Valley. The climatic conclusions of Fairbridge (1962, 1963) and of Williams & Adamson (1974) seem more accordant with the palaeolimnological evidence than those of some other investigators.

Palaeolimnological study in Africa has only begun. The Nile is likely to become palaeohydrologically, as it already is hydrologically, one of the best known rivers of the world.

Acknowledgements

I am indebted to the National Science Foundation of the United States for financial support, and to Mrs. Jennifer Angyal, Mr. Stephen Holdship, and Mr. George Sugihara for critical reading of the manuscript.

REFERENCES

Berry, L. & Whiteman, A. J. 1968. The Nile in the Sudan. Geogr. Jour. 134: 1–37.

Bishop, W. W. 1965. Hominidae and controversy. Geogr. Jour. 131: 254–256.

Bonnefille, R. 1970. Premiers résultats concernant l'analyse pollinique d'échantillons du Pléistocène de l'Omo (Ethiopie). C. R. Acad. Sc. Paris 270: 2430–2433.

Bonnefille, R. 1972a. Associations Polliniques Actuelles et Quaternaires en Ethiopie. Thèse, Univ. de Paris VI. CNRS A 07229 T. 1, 513 p.

Bonnefille, R. 1972b. Considérations sur la composition d'une microflore pollinique des formations Plio-Pléistocène de la basse vallée de l'Omo (Ethiopie), pp. 22–27. In E. M. van Zinderen Bakker (ed.), Palaeoecology of Africa, V. 7. A. A. Balkema, Cape Town.

Butzer, K. W. 1971. The lower Omo basin: geology, fauna and hominids of Plio-Pleistocene formations. Naturwissenschaften 58: 7–16.

Butzer, K. W., Brown, F. H. & Thurber, D. L. 1969. Horizontal sediments of the lower Omo valley: the Kibish formation. Quaternaria XI: 15–29.

Butzer, K. W., Issac, G. L. Richardson J. L. & Washbourn-Kamau, C. 1972. Radiocarbon dating of East African lake levels. Science 175: 1069–1076.

Clark, J. D. & van Zinderen Bakker, E. M. 1964. Prehistoric culture and Pleistocene vegetation at the Kalambo Falls, Northern Rhodesia. Nature 201: 971–975.

Coetzee, J. A. 1967. Pollen analytical studies in East and Southern Africa, xi + 146 p. In E. M. van Zinderen Bakker (ed.), Palaeoecology of Africa, V. 3. A. A. Balkema, Cape Town.

Cole, G. H. 1965. Recent archaeological work in southern Uganda. Uganda Jour. 29: 149–161.

Cole, G. H. 1967. The later Acheulian and Sangoan of southern Uganda, pp. 484–526. In W. W. Bishop & J. D. Clark (eds.), Background to Evolution in Africa. University of Chicago Press.

Fairbridge, R. W. 1962. New radiocarbon dates of Nile sediments. Nature 196: 108–110.

Fairbridge, R. W. 1963. Nile sedimentation above Wadi Halfa during the last 20,000 years. Kush 11: 96–107.

Glover, J. & Kenworthy, J. M. 1957. Effect of altitude on the temperature and dew point of the air. Rep. E. Afr. Agric. For. Res. Org. 18 p.

Greenwood, P. H. 1964. Explosive speciation in African lakes. Proc. R. Instn. 40: 256–269.

Greenwood, P. H. 1965. The cichlid fish of Lake Nabugabo, Uganda. Bull. Br. Mus. Nat. Hist., Zool. 12: 315–357.

Greenwood, P. H. 1974. The Cichlid fishes of L. Victoria; East Africa; The biology and evolution of a species flock. Bull. Br. Mus. N. H. Zoology Suppl. 6: 1–134.

Hamilton, A. C. 1972. The interpretation of pollen diagrams from highland Uganda, pp. 45–149. In E. M. van Zinderen Bakker (ed.), Palaeoecology of Africa, V. 7. A. A. Balkema, Cape Town.

Hecky, R. E. & Degens, E. T. 1973. Late Pleistocene-Holocene chemical stratigraphy and paleolimnology of the Rift Valley lakes of Central Africa. Woods Hole Oceanographic Institution Technical Report 73–28. 114 p. Unpublished manuscript.

Hedberg, O. 1951. Vegetation belts of the East African mountains. Svensk. Bot. Tidskr. 45: 140–202.

Hedberg, O. 1954. A pollen-analytical reconnaissance in tropical East Africa. Oikos 5: 137–166.

Hurst, H. E. 1952. The Nile. Constable, London. xv + 331 p.

Kendall, R. L. 1969. An ecological history of the Lake Victoria Basin. Ecol. Monogr. 39: 121–176.

Kilham, P. 1971. A hypothesis concerning silica and the freshwater planktonic diatoms. Limnol. Oceanogr. 16: 10–18.

Langbein, W. B. & Schumm, S. A. 1958. Yield of sediment in relation to mean annual precipitation. Amer. Geophys. Union Trans. 39: 1076–1084.

Langdale-Brown, I., Osmaston, H. A. & Wilson, J. G. 1964. The Vegetation of Uganda and Its Bearing on Land Use. The Govt. of Uganda, Entebbe. 159 p.

Livingstone, D. A. 1967. Postglacial vegetation in the Ruwenzori Mountains in equatorial Africa. Ecol. Monogr. 37: 25–52.

Morrison, M. E. S. 1961. Pollen analysis in Uganda. Nature 190: 483–486.

Morrison, M. E. S. 1966. Low-latitude vegetation history with special reference to Africa. Royal Meteorol. Soc. Proc. Int. Symp. on World Climate from 8,000 to 0 B.C. 142–148.

Morrison, M. E. S. 1968. Vegetation and climate in the upland of south-western Uganda during the later Pleistocene period. I. Muchoya Swamp, Kigezi District. J. Ecol. 56: 363–384.

Morrison, M. E. S. & Hamilton, A. C. 1974. Vegetation and climate in the uplands of south-western Uganda during the later Pleistocene period. II. Forest clearance and other vegetational changes in the Rukiga highlands during the past 8,000 years. J. Ecol. 62: 1–32.

Osmaston, H. A. 1958. Pollen Analysis in the Study of the Past Vegetation and Climate of Ruwenzori and Its Neighbourhood. B.Sc. Thesis, Oxford. 39 p.

Osmaston, H. A. 1965. The Past and Present Climate and Vegetation of Ruwenzori and Its Neighbourhood. D. Phil. Thesis, Oxford, n.p.

Saggerson, E. P. 1952. Geology of the Kisumu District, Degree Sheet 41, N.E. Quadrant. Geol. Surv. of Kenya, Report no. 21. 86 p.

Said, R. 1962. The Geology of Egypt. Elsevier, Amsterdam. xvii + 377 p.

Stuiver, M., Deevey, E. S. & Gralensky, L. J. 1960. Yale natural radiocarbon measurements, V. Radiocarbon 2: 49–61.

Temple, P. H. 1964. Evidence of lake-level changes from the northern shoreline of Lake Victoria, Uganda, pp. 31–56. In R. W. & R. M. Prothero (eds.), Geographers and the Tropics: Liverpool Essays. Longmans, London.

van Zinderen Bakker, E. M. 1962. A pollen diagram from equatorial Africa, Cherangani, Kenya. Geol. en Mijnbouw 43: 123–128.

van Zinderen Bakker, E. M. 1964. A pollen diagram from equatorial Africa. Cherangani, Kenya. Geol. en Mijnbouw 43: 123–128.

van Zinderen Bakker, E. M. & Coetzee, J. A. 1972. A re-appraisal of late-Quaternary climatic evidence from tropical Africa, pp. 151–181. In E. M. van Zinderen Bakker (ed.), Palaeoecology of Africa, V. 7. A. A. Balkema, Cape Town.

Williams, M. A. J. & Adamson, D. A. 1974. Late Pleistocene desiccation along the White Nile. Nature 248: 584–586.

30

II. THE NILE, A LIFE ARTERY

4a. PALAEO-ECOLOGY

by

J. RZÓSKA et al.

Egypt

Climatic changes in the past in Egypt and Nubia have been referred to in chapter 2 and are taken up in greater detail in the subsequent pages on the Sudan. Sediments, human camp sites, including traces of vegetation and cultivation and food animals have been used to build up a picture of the biological past along the Nile Valley. Another source of possible information is contained in rock drawings. With some discretion these can be taken as man's reaction to his environment and the fauna present. Winkler (1938, 1939) has published two volumes of 'Rockdrawings of Southern Upper Egypt'. These were collected in two seasons of explorations (1936–1938) in the Eastern Desert at the latitude of Quena-Luxor, on the eastern and western side of the Nile Valley opposite Luxor, at Silwa, and at Hosh and Aswah. In the Western Desert, the Kharga and Dakhla oases and Gebel Uweinat supplied further material. Dunbar (1930) has collected some in Egyptian Nubia. Winkler has tried to allocate the signs, inscriptions and pictures, often superimposed, to a chronological sequence and to interpret the human development. He distinguishes several phases of these drawings from the oldest, made by the earliest hunters (probably of the late Palaeolithic, to the mainly tribal signs made by Arabs in the last 1,000 and more years. For our purposes only some main points relating to the fauna, as biological indicator, are mentioned.

1. Faunal representations decrease with time advancing towards the present. They are numerous in the early phases where Winkler distinguishes four main groups of people as to occupation or distribution – 'Earliest Hunters, Early Nile Dwellers, Eastern Invaders, Autochthonous Mountain People'.

2. The earliest pictures show an intense preoccupation with the big African fauna; elephant and giraffe prevail, obviously as desirable objects of hunting, antelope, gazelles, Barbary sheep, wild cattle, wild ass, ostrich are frequent, some feline predators (lion and leopard) appear, and other species. Nile dwellers depict also the hippo, crocodiles and the Nile monitor (?). There must have been a savanna type of environment to sustain this fauna which is widespread over the whole area investigated.

3. Surprising is the profusion of wild fauna in the eastern desert along the wadis leading to the Nile, richer and denser than west of the Nile. Apparently more favourable conditions existed here and more vegetation

is still there at present. This difference between the two deserts has been stressed by a number of authors. McBurney (1960) says: 'Not the least remarkable of the Nile's characters is the suddenness of the way in which it delimits the Libyan desert'. To the west extends a mature desert with no fluviatile remains, to the east of the Nile remnants of deep well preserved valleys exist, wells are numerous and there is some desert pasturage. Water supply in the early stages of the Nile system came from east and the Red Sea hills. A geological map (R. Said 1962) shows these wadis, remnants of former rivers or drainage channels. The map of the Nile in ch. 8 and the space photograph of the Lake Nasser basin in ch. 19, show the mouths of these wadis, now drowned.

4. The gradual disappearance of the big fauna is well marked; first the elephant and later the giraffe vanish from the rock drawings, and later again other species. But the succeeding pastoral people have left a rich

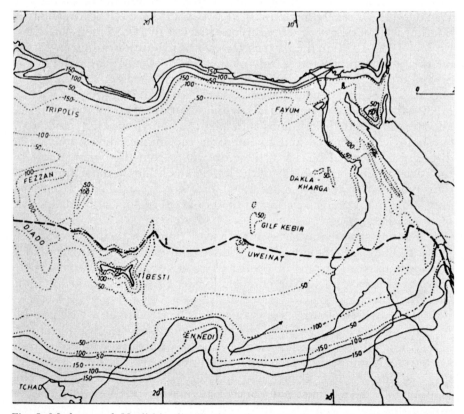

Fig. 5. Modern and Neolithic distribution of precipitations in the Eastern Sahara; full lines: present isohyets; dotted lines: reconstructed isohyets from sixth to fourth millenium B.C.; dashed line: present boundary between predominating winter and summer rains; from Butzer 1966.

34

documentation of their domesticated cattle in wadis with vegetation and wells, with fewer references to wild life. Cultivators follow with settled habits on oases, some of these now extinct. Faunal rock pictures from dynastic times contain antelope, gazelles, lion and birds (sand grouse?); the aquatic environment is represented well, with hippo (an astonishingly 'modern' sketch is recorded in Vol. II, plate X, 2), crocodile and fish. The profusion of aquatic and marsh birds on bas-reliefs and wall paintings has already been mentioned. Pictures of the elephant become rare, but see below. Graeco-Roman-Coptic inscriptions and those of Arab times contain little or nothing about the fauna. No dating could be done by Winkler but remarks to the 'Badarian, Amratian, Gerzean' cultures and those of the groups 'A, B and C' enable us to locate these phases to late Palaeolithic, Neolithic, Predynastic and Early Dynastic times.

Butzer (1971) recognises some climatic changes in the last 30,000 years in Egypt. A wet phase ending about 25,000 BP; a dry phase 24–18,000 BP but wetter in the sub-Saharan drainage basin of the Egyptian Nile; from 17,000–5,000 BP, a phase with winter rains in Egypt, a greater Ethiopian discharge and a northward shift (Fig. 5) of summer rains in the Sudan; from 10,200 to 8,000 and 6,000 to 5,000 BP there were some sub-pluvial 'episodes'. These moister phases, in a gradually drier climate, were able to sustain vegetation along the wadis. 'Profusions of terrestrial snails, now extinct in Egypt, further suggest a moister climate. The end of this moist phase coincides with the extinction of elephant, rhinoceros, giraffe and gerenuk gazelle between the 1st and the 4th Dynasties, i.e. between about 2,900 and 2,600 BC'. However, reliefs of the 5th Dynasty still show wild animals in open landscape 'studded with tree size sycamores and acacias as well as desert shrubs'; a desert grass savanna persisted still, especially in the wadis and along the desert margin. Later hunting scenes show a much poorer environment (Butzer 1966). Independently, Arkell (1961) recognises a wetter interlude at 6,000 BP. According to him, even during the first dynasties – about 3,100 BC-, the southernmost administrative district (nome) of Egypt was called 'Elephant-nome'; other animals were adopted for more northerly nomes. It is at this time that the ultimate phase of desiccation began to be felt more decisively. By the Middle Kingdom some of the big African mammals were only known by hearsay and became symbols and hieroglyphic signs of occupations and prowess. The process of retreat of the African fauna was a consequence of the growing spread of desert conditions; man's role in this process was at that time probably insignificant.

'The mid-Holocene moist interval appears to have been terminated completely about 2350 BC, after a first climatic crisis had been sustained five or six centuries earlier. During and after the close of the 6th Dynasty (c. 2350–2180 BC) the Nile floods no longer inundated the peripheries of the flood plain, and blowing sands from the Libyan desert covered much of the former alluvium . . . the implications can be seen from various famines recorded between 2,100 and 1,950 BC and were attributed to failure of the annual floods' (Butzer 1966).

Kees (1961) in his 'Cultural Topography of Ancient Egypt' adds to the knowledge of Egyptian palaeo-ecology. During the Early Stone Age the Nile Valley north of Nubia was marshy and 'uninhabitable'. The bed of the Nile in Upper Egypt lay at a higher level than today, about 5 m at Luxor, and its waters were damned by the rock barrier of Gebel Silsilah, forming a lake which stretched as far as the cataract of present Aswan. North of Thebes the country at that time resembled the 'Sudd' of the southern Sudan.

Special mention must be made of the Faiyum, besides the Delta the most conspicuous part of Egypt, and remarkable for its archaeological significance. Andrews (1906) gave a list of Tertiary vertebrates found in Middle and Upper Eocene deposits; these are totally different from the present fauna and included marine forms; Miocene and Pliocene formations followed, finally ending in the Pleistocene with elephant, hippopotamus, *Bubalis*, *Canis* and crocodiles, Chelonia and fishes; stumps of trees suggested woodland. This African fauna persisted here longer than in other parts according to Kees.

A site investigated on the northern shore of the Faiyum lake yielded industries closely associated with riverine or lacustrine environments of upper palaeolithic character and contained enormous quantities of fish bones, evidence of fishing economy; dates of some of these sites are about 6,000 BC.

At Wadi Natrun in the Egyptian desert, where recently an isolated colony of Papyrus was discovered, a rich fauna of nilotic fishes existed in the Pliocene (Greenwood 1972); 15 freshwater and 4 marine genera could be identified including: *Protopterus*, *Polypterus*, *Hyperopisus*, *Hydrocynus*, *Alestes deserti* n.sp., *Labeo*, *Barbus*, *Clarotes*, *Bagrus* (?), *Auchenoglanis*, *Synodontis*, *Clarias*, *Heterobranchus* (?), *Lates*, *Tilapia*, *Chrysophysis*. This is about one-third of the total 48 genera of Nile fishes. The marine fishes are probably euryhaline and their presence indicates a connection of the freshwater lake existing there with the sea.

Both Winkler and Kees remark on the profusion of boat drawings from prehistoric time onwards. This is understandable, even now the Nile is a much used route of communication – in the past it was the only one. Most of the rock drawings represent timber boats and, like in recent times in the Sudan, this may have contributed to the disappearance of trees, together with overgrazing and destruction by the big herds of cattle (Kees).

This brings us to historic times. The landscape, flora and agriculture of Ancient Egypt is described by V. Täckholm (ch. 4c.). In the Delta with perennial swamps and inundations, conditions favouring cattle herding persisted for a long time.

Food producing started in Upper Egypt and the Faiyum; the Delta ('the land of papyrus'), although a 'great reserve of land for internal colonisation' was accessible only after centuries of canal building and dykes in the 10th–6th century BC. One important factor in human economy was the early development of river traffic. The Nile was a link that united the country and remained so to the 19th century AD as the main gateway into Africa. But this long narrow valley of the Nile hemmed in on both sides by increasing desert conditions created also environmental, racial and economic differences. Fluviatile civilizations arose along the Nile in Nubia; the Meroitic phase was followed by a period of christian culture from the 6th to the 15th century and then the Arab conquest.

Sudan

The physiography of the Sudan has undergone considerable changes in the past and some of the features have already been mentioned in a previous chapter 2. These relate mainly to the Nile valley itself. But the landscape surrounding a river has great influence on the valley and a general assessment of the terrestrial ecology is necessary.

In order to appreciate fully the changes, the present state of the climate and the Sudanese landscape must be brought into view in this chapter. The best indicator of ecological conditions is the vegetation described in chapter 5, based on sources supplied by G. Wickens (1975). He also kindly perused part of my presentation. The vegetation map reproduced later (Fig. 10) shows the enormous extent of desert and semi-desert conditions, occupying at present half of the country. Detailed explanations to the map characterize the main vegetation components. A clear relationship exists between the climatic features; rainfall belts (Fig. 11) run in a regular fashion across the country with few enclaves and form the basis for the distribution of plants and animals.

The present arid and hot conditions developed relatively late within the last 20,000 years (Fig. 6). Wickens (1975) has adapted the timetable of events proposed by Warren (1970) as follows:

20,000–15,000 BP	Very Arid: rainfall belt moved 450 km southwards, desert conditions advanced, sands reached 10° lat. N.
12,000–7,000	Wet; White Nile lakes south of Khartoum were formed, probably additionally impounded by sediments of Blue Nile at junction; lake conditions extended up to 500 km upstream; according to Tothill (1948) 'Sudd' deposits reach northwards to below southern margins of the great lake postulated by the later research. Rainbelts shifted towards the north for about 400 km, with rainfall in the central Sudan 2/3 times greater than at present. In the volcanic outcrop of Gebel Marra, a number of plants have been discovered, now outside their present range (Wickens, G. 1975 and also monograph on G. Marra to appear 1975); similarly in the Malha crater north of Gebel Marra.
7,000–6,000	Fluctuating periods; shifts of rains south and northward retreat of wind pattern, advance of dune colonising plants to the north.

37

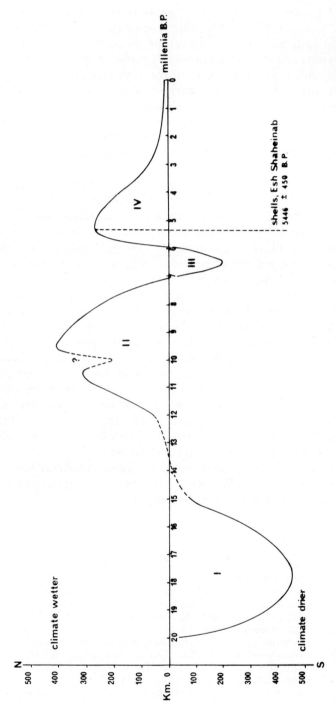

Fig. 6. Climatic shifts in the Sudan from 20,000 B.P. to Present; Note: the shape of the curve for each climatic period is not accurately known; G. Wickens (1975).

	There is at present no biological evidence to support Warren for this phase.
6,000–3,000	Moderately wet, shift of climatic belts north, riverine swamps at the epi-palaeolithic site at 'early Khartoum' 7,000–8,000 B.P., but not at the early Neolithic site of Esh Shaheinab, 5,000–5,500 B.P.
3,000–recent	Climate becomes gradually drier until today's conditions of super-aridity in the northern part of the Sudan are reached.

Against this background some biological developments took place; first is the evidence derived from excavations in Sudanese Nubia by Wendorf *et al.* (1968). Man's presence in prehistoric Nubia is attested by numerous camping sites with stone implements of various techniques (industries), classified by prehistorians from Acheulean, Mousterian, both dated before 45 and 33 thousand years BC respectively, down to the Neolithic; these were followed by Nubian Ceramic predynastic and historical phases. The impression is gained of many separate groups, mostly small but growing in size with time, which settled for short or longer times near or in the Nile Valley. This Nubian Nilotic culture, demanding adjustment to the specific conditions of the Nile valley, begins to emerge in a more definite form with the establishment of the present regime of the Nile. The ethnic origin of these human populations is at present still uncertain; burials found in Qadan sites, dated 9,460 B.C. $\pm$ 440 point to a Cro-Magnon type with affinities to the type found at Mehta in the Maghreb. Contacts across the present Sahara, not a desert barrier then, seem to be indicated by similar affinities in other sites e.g. an Arkin site dated 7,440 B.C. $\pm$ 180.

As in prehistoric Egypt, people in Sudanese Nubia derived their livelihood from the earliest observed times down to about 5,000 B.C. from hunting with fishing in second place. The astonishing discovery of grinding stones, 10,000 B.C. and possibly earlier, points to another food source; this new discovery is further amplified by 'lunate' stone tools with 'lustrous' edges, probably hafted on sticks and used as sickles, found at the Arkin site named above.

VEGETATION

Vegetation was certainly quite different from the present desert conditions. But the composition of the vegetation is unknown; pollen has not been found in Nile sediments so far except in 'ponds' created by floods or seepage near Esna in Upper Egypt about 10,000–9,000 B.C. There large cereal type pollen, together with algal deposits, have been identified tentatively as wild barley. Reeds of some kind grew also in these pools. But the best testimony for the vegetation comes from the faunal remains found in association with human settlements near the Nile. These clearly indicate a grassland savanna able to support a varied mammal fauna.

39

According to A. Gautier (in Wendorf (ed.) 1968 p. 88 ff.) the following could be identified in Nubia from bone remains: *Lepus* sp. (probably *L. capensis*), *Canis aureus* cf. *sudanicus* still extant, *C. mesomelas* now found only much further south, *Equus asinus cf. africanus* now probably extinct, *Hippopotamus amphibius*, *Bos primigenius* still hunted during dynastic time, *Adenota kob*, of which 3 ssp. still occur in the Sudan, *Alcelaphus buselaphus* (hartebeest) still extant in the Sudan, *Gazella dorcas*, *G. rufifrons* and *G.* sp., the two first living still, *Capra hirca*, now extremely rare in isolated spots.

Of these the most hunted in Nubian Nile sites were *Bos*, *Alcelaphus* and *Gazella rufifrons*; the hippopotamus and the wild ass appear less frequently, others only sporadically in the bone remains. We will note here the largely African character of this fauna as testimony of the link between the lower Nile valley and the rest of the continent; this link is treated in a separate chapter (ch. 6). Further evidence is provided by rock drawings (see above).

Aquatic fauna

Fishes appear in camp sites less frequently and the remains could be identified only in some cases. It is possibly significant that no fish remains have been found in the earliest site, except the Faiyum, but appear after the establishment of the present Nile regime. Seven genera could be identified by P. H. Greenwood (in Wendorf (ed.) 1968 p. 100 ff.): *Clarias*, *Bagrus*, *Chrysichthys?*, *Synodontis*, *Barbus* (cf. *bynni*), *Tilapia*, and *Lates*. *Clarias*, a catfish with additional breathing organs, is represented at all sites even in the oldest over 20,000 B.C., the others less frequently. All are of Nilotic character (see chapter on the Nile fish fauna by P. H. Green-wood), but many elements of that fauna are absent in Nubian sites. The fish remains are scanty, in many cases unidentifiable; their presence at human sites may be selective and no conclusions or wider nature can be drawn on the basis of this material.

Mollusca

Mollusca are richly represented as one would expect from the resistance of their shells. Francine Martin (in Wendorf ed. 1968 p. 53 ff.) found 14 species of snails and 10 of lamellibranchs in the Wadi Halfa region. Of these 10 appeared only in the fossil state, 6 only modern (at present in the Nile there); both fossil and present totalled 8 species. The detailed species list is:

Theodoxus niloticus, not fossil, but abundant now in Nile, also in Ethiopia, palaearctic.
Viviparus unicolor, only fossil, now in East African lakes and Bahr el Ghazal.

Lanistes carinatus, only fossil, otherwise widespread in Nile.
Valvata nilotica, fossil and modern, palaearctic affinity.
Bithynia neumanni, only modern, widespread in East-Central Africa incl. Chad, fossil in
  Faiyum.
Gabbia walleri, only modern, now only in Lake Albert.
Melanoides tuberculata, only modern, widespread in Egypt and Sudan, fossil in Egypt.
Cleopatra bulimoides, both fossil and modern, widespread in Nile.
Bulinus truncatus, fossil and modern, palaearctic affinity, Africa.
Anisus planorbis, only fossil, palaearctic.
Gyraulis costulatus, fossil and modern, widespread in Africa.
Farissia sp., fossil and modern, North Africa.
Zootecus insularis, only fossil, N. Africa, Ethiopia, Somalia, Arabia, India, dry con-
  ditions.
Caelatura aegyptiaca, fossil and modern nilotic.
Unio abyssinicus, only fossil, in all older sediments, disappeared after the Arkin phase,
  fossil in Faiyum.
Apatharia hartmanni, only fossil, now lives in Upper Nile, Ethiopia, East African lakes.
A. cailliaudi, fossil only at Wadi Halfa, but widespread in Nile, Ethiopia.
Mutela nilotica, fossil and modern.
M. emini, only fossil, now in Central Africa.
M. cf rostrata, fossil only, now Equatorial Africa and Niger-Chad.
Etheria ellyptica, both fossil and modern, Nile, Niger, Congo (Zaire).
Corbicula consobrina, only modern.
C. yara, only fossil.
Sphaerium hartmanni, only modern.
(*S. hartmanni cf. mohasicum*, variety of above, only modern, very near to East African
  lakes specimens).

Some species have specific requirements and specialists have suggested that the absence of *Unio willcocksi* (= *U. abyssinicus*) in the modern Nile is due to the present higher water temperatures; it lives at present in Lake Tana.

Further south, reliable indicators of palaeo-ecological conditions are molluscs, especially land and amphibious snail species. The distribution of the land snail, *Limicolaria cailliaudi* (L. flammata) is significant and can be regarded as supporting rainfall shifts. At present its northern limit is the 400 mm isohyet, from Sennar to Kosti, but found sub fossil at the Khartoum Epi-palaeolithic site and also the Esh Shaheinab Neolithic site.

Wickens (1975) has reconstructed vegetational maps which would follow the shifts of rain belts causing increase of humidity when moving northwards and aridity if they moved south. Such shifts would not only affect the distribution of plants, animals and man, but also the activation of water courses now extinct. An example is the Wadi Howar, ca. 16 °N, 24 °30′E which opens to the Nile between the IV and III cataract. R. Said (1975) regarded the Wadi Howar and the near Wadi Milik as potential sources of previous phases of the Nile e.g. in the Upper Pliocene and interesting observations on the present state of the Wadi Howar may support its ancient role. Wickens (1975) has collected the evidence. At least for some months there is still sufficient water to support a number of mammals (Shaw 1935). The list includes three gazelles, Addax-Dorcas-Addra, the white Oryx, the Barbary sheep, giraffe, the red Hussar monkey, ant bear, porcupine, jerboa, gerbil, hyrax, hare, lion, African hunting dog, striped hyaena, jackal, Fennec fox and two uncertain species (cheetah and Civet cat). At present the area receives 150 mm. rainfall per year. At Bueira further 'upstream' there is even sufficient moisture for the survival of the amphibious snail, *Pila (Ampullaria) wernei*. Although large numbers of the snail had perished either from desiccation or by predators, some had been able to survive in sheltered places (Arkell 1945) . . . The area receives today about 230 mm. rainfall (Ramsay 1958) . . . and is well outside their usual distribution range . . . normally upwards of 400 mm. (Tothill 1946). Today there are few sources of permanent water in the Wadi Howar . . . The available evidence suggests that a very little increase in the rainfall would have enabled neolithic man to have survived in the Wadi Howar area.' This applies obviously to the whole area across the Sudan; with the shifts postulated, ecological changes would occur, with the delay obviously necessary for the vegetation to respond.

Camp sites of prehistoric man south of Nubia provide further evidence of former spread of animals and plants. But Wickens (1975) restricts their value in some degree: they are all riverain (e.g. Khartoum epi-palaeolithic and Esh Shaheinab), vegetation is always more flourishing along rivers and thus would favour the presence of mammals both as corridors of movement and habitat. Also, species differ in their value as indicators of ecological conditions; the Nile Lechwe is clearly a swamp animal and so is (was) the reedrat of the Khartoum epi-palaeolithic site, Oryx and Addax are semi-to-arid zone species. But others, e.g. giraffe and elephant migrate in search of food.

Some details of faunal presence from well known prehistoric sites may be mentioned.

Singa and Abu Hugar on the Blue Nile, dated 17,500 BP but probably much older, yielded the famous Singa skull together with some species now extinct; also the hippo and rhino indicating a semi-dry landscape with riverain influence (Bate 1951).

Khartoum, 8,000 BP, now named as epi-palaeolithic, had a rich assembly of animal remains with the reed-rat, now extinct; all other species are still existing much further south, hippo, warthog, Nile lechwe, buffalo, black rhino, elephant, a number of gazelles and the leopard. This was a riverain swamp surrounded by savanna (Arkell 1949).

Shabona on the White Nile south of Khartoum, 7,000–8,000 BP, had a fauna similar and probably contemporaneous with Arkell's 'Early Khartoum'; again a river swamp and savanna (Clark 1973).

Shaheinab, north of Khartoum, 5,440–5,000 BP, had remains of monkey, lion, leopard, giraffe, rhinogazelles and antelopes, but not the reed-rat nor the Nile lechwe.

42

Gebel Tomat, White Nile south of Khartoum, Yielded only small vertebrates of savanna background (Clark 1973).

This great African fauna was widespread in northern Africa including the Sahara. The famous rock-drawings of Tassili el Ajjer described by Tschudi (1956), Lhote (1958) have been recently interpreted and put into some historic sequence by Hugot (1974); they correlate closely with those found in Upper Egypt and the Gebel Uweinat. Nilotic fresh water molluscs have been found in mid-Sahara (Sparks & Grove 1961) (see Fig. 12.).

Although Berry (ch. 2) has discussed the Nile system in the Sudan, some remarks may be added on the palaeolimnology of the two rivers in this area. The evidence for this comes from investigations of soil conditions initiated by the Ministry of Agriculture of the Sudan, and carried out by Hunting Technical Services and other commercial firms; it was extended scientifically by members of Macquarie University, Australia (Williams, Adamson), by J. D. Clark and an archaeological team from Berkeley University, U.S.A. with the help of local scientists. Tothill (1946, 1948) had already recognised the importance of mollusc shells for explaining the origin of the great fertile plain of the 'Gezira' which stretches across the angle of two rivers. An area of about 1,250 km² forms the main agricultural centre of the Sudan with irrigated cotton cultivation mainly, and from this stemmed the subsequent investigations.

Williams & Adamson (1974) have established a sequence of events in the Sudan starting from a very dry phase all over East Africa and as far as Lake Chad in the north; this lasted for several thousands of years and ended about 12,000 BP. The White Nile was greatly reduced during that period and earlier (30,000 BP); the authors write: 'The White Nile was deprived of water from the Uganda lakes, so that its flow dwindled to a mere seasonal trickle . . . deposits of very pure carbonate accumulated in or along the river, which contained abundant calcium and magnesium, indicative of the high salinity concentrations brought about by increasing evaporation downstream – a feature of modern rivers in arid areas'. (A site on the east bank north of Kosti) . . . 'contained both dolomite and calcite crystals, suggesting that when the dolomite was forming magnesium-rich calcite was already abundant, and the ratio of dissolved Mg to Ca was at least 7'.

At about 12,000 BP (see ch. 3) the White Nile rose to 3–7 m above the high level of today. A lake was formed up to 20 km broad with smaller lakes 15 km west of the present shores. These were recognised by ample mollusc shells, containing all the species known at present to be alive (Williams 1966, Williams & Adamson 1973). The rainfall in this phase, lasting from 12,000 to about 7,500 BP, was much higher than at present with a savanna vegetation; this coincided with the 'Early Khartoum' (Arkell 1949) settlement, now regarded as epipalaeolithic. After 7,500 BP the White Nile began to fall, dark clay deposits from the swamps fringing the receding river with amphibious molluscs. The gradual desiccation continued into the Neolithic; the site at Shaheinab about 5,000 BP (Arkell, 1953) shows clearly the impoverishment in flora and fauna. A neolithic site on the east bank of the White Nile north of Kosti is worth mentioning specifically (Williams & Adamson 1973); potsherds found there contained a large admixture of sponge spicules of *Eunapia nitens* Carter which was found in 1956 in large colonies on reeds near Kosti harbour (unpublished); the practice of admixing sponge spicules is apparently known also in parts of America.

The Blue Nile about 7,000 BP was 4–5 m. higher than at present, attested again by terraces with shells. This is the river which created over several thousands of years the Gezira plain as an 'alluvial fan'. Large inundations and river flows streamed across this plain about 6,000 BP, westwards towards the White Nile at the time when this river dwindled. The sediment load must have been great; at Roseires deposits of 60 m exist and are now being incised by the present river. This sequence agrees in broad outlines with that of Warren and Wickens. For the Main Nile north of Khartoum see the chapter by Berry.

Uganda

The East African Lake Plateau provides the head waters of the White Nile system, a landscape quite different from the Sudan and Egypt. The great variety of surface forms, and their origin, the existance of large lakes with a remarkable fauna and the discovery of some of the earliest sites of hominid evolution has been the subject of many investigations and a very large literature.

In this book the geological and climatic sequences around the head-waters of the Nile and their origin are treated by Livingstone (ch. 3). Greenwood (ch. 7) traces the history and extent of the fish fauna as testimony of the past. Thompson (ch. 11) describes the former and present lake swamps. Climatic data have been referred to in the chapter on Lake Victoria (ch. 10). Some very significant rainfall diagrams have been used in Fig. 11 to characterise differences along the Nile valley. For Ethiopia see ch. 2, 3 and 14.

In conclusion we need only to say that in the last 20,000 years conditions have been on the whole fluctuating with richer conditions of life throughout the environment of the Nile until the last present phase of desiccation, sharply brought to general awareness by the recent drought in the Sahel. The grip of the desert applied most strongly to the middle and lower courses of the Nile valley, where conditions have declined severely. The enormous area between Khartoum and the Egyptian Delta shrank as habitat for life to the thin line of the Nile valley, and even here with interruptions.

REFERENCES

Andrews, C. W. 1906. The extinct animals of Egypt. A descriptive catalogue of the Tertiary Vertebrates of Faiyum. Brit. Museum Nat. Hist.

Arkell, A. J. 1945. Some land and freshwater snails of the western Sudan. Sudan Notes and Records 26: 339–341.

Arkell, A. J. 1949. Early Khartoum. Oxford University Press, London.

Arkell, A. J. 1953. Shaheinab. Oxford University Press, London.

Arkell, A. J. 1961. A history of the Sudan until 1821. Athlone Press, Univ. of London, second edition.

Bate, D. M. A. 1951. Mammals from Singa and Abu Hugar. The Pleistocene fauna of two Blue Nile sites. Fossil Mammals of Africa no. 2, Brit. Mus. Nat. Hist.

Butzer, K. W. 1966. Climatic changes in the arid zones of Africa during early to mid-Holocene times. Roy. Meteor. Soc.: World Climate from 8,000 to 0 BC. Proc. 72–83.

Butzer, K. W. 1971. Environment and Archaeology, second edition. Methuen & Co.; Aldine Atherton, Chicago/New York.

Butzer, K. W. 1972. Radio-active dating of East African lake levels. Science 175: 1069–1075.

Clark, J. D. 1973. Preliminary report of an archaeological and geomorphological survey in the Central Sudan, January–March 1973. Mimeo.

Dunbar, J. 1930. Rockdrawings of lower Nubia. Service des Antiquités. Le Caire.

Greenwood, P. H. 1972. New fish fossils from the Pliocene of Wadi Natrun, Egypt. J. Zool. Soc. London 168: 503–519.

Hugot, M. M. 1974. La Sahara avant le désert. Editions des Hespénides, Toulouse.

Kees, H. 1961. Ancient Egypt: A cultural Topography, (ed.) T. G. James. Faber & Faber, London.

Lhote, H. 1958. À la découverte des fresques du Tassili. Arthaud, Paris.

McBurney, C. B. M. 1960. The Stone Age of Northern Africa. Penguin edition.

Said, R. 1962, 1975. See chapter 1 and 2.

Shaw, T. B. K. 1936. An expedition to the southern Libyan desert. Geogr. J. 87: 193–211.

Sparks, B. W. & Grove A. T. 1961. Some quaternary fossil non-marine Mollusca from the Central Sahara. J. Linn. Soc. London 44: 355–364.

Tothill, J. D. 1945. The origin of the Sudan Gezira clay plain. Sudan Notes and Records 27: 153–183.

Tothill, J. D. 1948. (ed.) Agriculture in the Sudan. Oxford Univ. Press, London.

Tschudi, Y. 1970. Les peintures rupestres du Tassili-el-Ajjer. À la Baconnière-Neuchatel.

Warren, A. 1970. Dune trends and their implications in the Central Sudan. Zeitschr. Geomorpholog. Suppl. 10: 154–180.

Wickens, G. (unpubl.). Palaeobiology. In: J. D. Clark et al. 'Prehistoric Settlement in the Central Sudan'.

Wickens, G. (1975). Changes in the climate and vegetation of the Sudan since 20,000 BP. Proc. 8th Plen. Session AETFAT, Genève 1974. Boissiera 24.

Wickens, G. 1975. Jebel Marra. The flora of Jebel Marra, Sudan Republic, and its geographical affinities. H. M. Stationary Office, Govt. Bookshop. (in press).

Williams, M. A. J. 1966. Age of alluvial clays in the western Gezira, Rep. of the Sudan. Nature 211: 270–271.

Williams, M. A. J. & Adamson, D. A. 1973. The physiography of the Central Sudan. Geograph. J. 139: 498–508.

Williams, M. A. J. & Adamson, D. A. 1974. Late Pleistocene Desiccation along the White Nile. Nature 248: 584–586.

Winkler, H. 1938–1939. Rock drawings of southern Upper Egypt, vol. 1, 2. Oxford University Press, for the Egypt Exploration Society.

4b. PREHISTORIC CIVILISATIONS IN LOWER EGYPT

by

M. S. GHALLAB

EDITOR. Prof. M. S. Ghallab has supplied a detailed chapter on the sequence of cultural development in Egypt. It has been found necessary to summarise this large contribution so as to fit it into the general context of this book. For this the editor is responsible.

The intense excavations in Nubia have been briefly described in ch. 1a. The sequence of cultural development in lower Egypt is attempted now. The Palaeolithic phase there is known only from a limited number of sites, due to the large deposition of sediments by the river northwards. Further levelling of the land by intense wind action has destroyed many early river and lacustrine terraces bordering the Nile.

The first palaeolithic site at Abbassia, east of Cairo, was discovered in 1920 by Bovier-Lapierre (1925). They were interpreted by Sandford & Arkell (1929) as Pre-Chellean, Chellean and mainly Achculean; according to Breuil (1931) some of the tools are of Levallois character. Similar tools were found at Wadi Tumilat, near the Abu-Soweir airport (Montet 1937) and at Heliopolis. A rich site of the late Palaeolithic was found at Kom Ombo by Vignard (1928) and named 'Sebilien' with Mousterien tools and a rich savanna fauna. This 'industry' was widespread along the Nile as mentioned in ch. 1. on Nubia. Similar sites were discovered at Silsileh, opposite to the mouth of the Wadi Kharit which at that time may have still been an active supply of water (Vignard 1955). Finally Vaufrey (1969) reported finds at Nag Hammadi, 170 km north of Kom Ombo.

The best sequence of cultural development comes from the Faiyum, the remarkable 'oasis' adjacent to the Nile. Many investigations have been carried out there of which those by Caton-Thompson & Gardner (1926, 1934, 1952) are famous. The Faiyum depression was the site of a sequence of lakes, with lacustrine terraces left behind, which have yielded rich depositories of biological and human remains. According to a new review by Wendorf & Schild (1975, in press) the sediments building up terraces allow to discern a succession of four lake phases: Palaeo-Moeris, Pre-Moeris, Proto-Moeris and the historical lake Moeris. Their different extent and depth reflect some heights of the levels of the Nile, which supplied the water. Only few traces have been found of the Palaeo-lake and these are still undated. Pre-Moeris left sediments at 17 m a.s.l. and two radio-carbon dates of about 6,000 B.C. The lake level fell to

12 m a.s.l. about 5,500 B.C. Subsequently the level rose to its highest mark of 24 m a.s.l. with one date of 5,190 ± 120 B.C. From about 4,000 B.C. the lake stood at a level of 23 m a.s.l. It was used as a reservoir for the overflow during flood periods; large parts of it were reclaimed for agricultural needs during dynastic times with ultimate disruption of the free flow from the Nile. Irrigation needs have been supplied for centuries by a canal, the Bahr el Yussuf. At present only a remnant of the former lake exists, the Birket El Qurun with increasingly brackish water due to evaporation. The rest of the Faiyum depression is intensively cultivated by irrigation. The 'oasis' is surrounded by desert.

An almost continuous sequence of human 'industries' from late Palaeolithic to dynastic times bears testimony for its favourable habitat in Pre- and Proto-Moeris times, from about 6,000 B.C. onwards. Stone implements, fish harpoons, bones of large mammals (including the hippo), and grinding stones testify to the early history. Some gaps exist until the full development of the Neolithic civilisation is reached.

These Neolithic cultures developed both in Faiyum and the Kharga Oasis and at a number of sites from the desert edge of the Delta upstream to Upper Egypt. The ecological conditions prevailing at that time have been described in ch. 4a. of this book. The increasing desiccation drove people towards permanent water. With changes in flora and fauna, hunting, fishing and cattle herding was replaced gradually by the first stages of agriculture and more settled habitations; grain stores, better built huts and ornamented pottery showed this advance. Distinct sequences of these aggregations have been given distinct names, which in the context of this book are not essential. The Pre-dynastic period starts about 4,000 B.C. Though there are no written records, yet this phase is well documented by pottery, burials and traces of permanent villages. The significance of rock drawings for the gradual changes has already been mentioned in ch. 4. Striking innovations appeared e.g. copper smelting, far in advance of its spread and use in Europe. Villages grew into larger district centres e.g. at Maadi, now a suburb of Cairo. A social structure emerged regionally with ensuing struggles for supremacy; power and centralisation led to the emergence of the historical age. A distinct Egyptian civilisation almost 'sprang' into being rapidly about 3,500 B.C. with a script, monumental stone buildings, elaborate graveyards, a set religious cult and a set social structure. All this evolved from cultures not different from those contemporary in other parts of the nearby world. Cultural affinities existed with parts of Africa and Asia with migrations and influences, including domestication of animals and plants; this is mentioned in ch. 4c. by V. Täckholm.

The decisive factor for the development of a distinct Egyptian civilisation and its continuity through thousands of years is the Nile. Egypt is a fluviatile civilisation and the crown of human development along this life artery.

REFERENCES

(EDITOR. From the large list given by M. S. Ghallab, only some have been selected).

Bovier-Lapierre, P. 1925. Stations préhistoriques des environs du Caire. Compte Rendu Congr. Intern. Géographie Cairo t. 4.

Breuil, H. 1931. L'Afrique préhistorique. Ed. Cahiers d'Art. Paris.

Caton-Thompson, G. & Gardner, E. W. 1926. Research in the Faiyum. Ann Egypt Part I.

Caton-Thompson, G. & Gardner, E. W. 1934. The Desert Faiyum. Roy. Anthrop. Inst. London.

Caton-Thompson, G. 1952. Kharga Oasis in Prehistory. University of London, Athlone Press.

Montet, A. 1937. Les industries Lavalloïsiens d'Heliopolis et d'Abu Souvain. Bull. Soc. Préh. Française. 54.

Sandford, K. S. & Arkell, W. J. 1929. Palaeolithic Man and the Nile-Faiyum Divide. Chicago Univ. Press.

Vaufrey, R. 1969. Prehistoire de l'Afrique. Université de Tunis.

Vignard, E. 1928. Une nouvelle industrie lithique, le Sebilien. Bull. Inst. Français, 25. Paris.

Vignard, E. 1955. Un Kjoekkemmonding sur la rive droite du Wadi Shait dans le nord de la plaine de Kom Ombo (Hte Egypte). Bull Soc. Préhist. Française 52.

Wendorf, F. & Schild, R. 1975. The Palaeolithic of the lower Nile Valley. Proc. Intern. Conf. North Afr. & Levantine. Pleïst. Hist. S. Methodist Univ., Dallas, Texas.

White mulberry, was introduced from India during the time of the early Caliphs together with the silk worm, which feeds on it.

Fruits of *Rosaceae*. Most fruit trees of this family, e.g. *Prunus*, *Malus*, *Pyrus*, *Cydonia*, were not known in Pharaonic Egypt, but introduced during the Graeco-Roman period. As a rule these trees require a colder climate and do not thrive very well in hot regions. In tombs of Graeco-Roman age at Hawara, Faiyum, were found kernels of Peach, Plum and Apple. Most probably also Apricot and Pear were introduced at that time, although there is no proof of it. Quitten, *Cydonia*, may have been the last introduction of the *Rosaceae*-group. There is no find known of it up till now.

Musa sapientum, Banana. There is no evidence of its cultivation in Pharaonic Egypt. The only record of it is a banana leaf from the 5th cent. A.D. found in the necropolis of Antinoë. Banana cultivation seems to have been established in Egypt during the Middle Ages. Prosper Alpinus, 1580, states that it is rather widespread around Damietta.

Citrus. Nowadays this genus represents the major part of fruits cultivated in Egypt. However, its cultivation in ancient Egypt seems to have been very restricted. Only three properly identified and dated finds are known, from Medinet Habu, 18th Dyn., from Antinoë, Greek period, and from the Monastery of Phoebammon, Thebes, early Coptic. Other records are either wrongly determined or lack dating.

All the three finds represent *Citrus medica*, Citron, as may have found its way into Egypt from South Himalaya through Persia, Media or Mesopotamia, at least to judge from its old name Median apple, Persian apple, etc.

The next Citrus to be introduced were the small-fruited Lime, *Citrus aurantifolia*, and the Bigarade, *Citrus aurantium*. Probably this took place during the 10th cent. A.D. through Oman or Palestine. All other species, as Orange, Lemon, Mandarine, Grapefruit, etc. are of late introduction.

Vitis vinifera, Vine. This has been cultivated since times immemorial. Small carbonised kernels from the first Dyn. were found in tombs at Abydos and Nagada. At that remote period, the wine-press was already used as a hieroglyphic sign, and wine-jars were abundantly present in the tombs. Vine-cultivation in Egypt is documented from all periods, whether depicted on tombs, recorded in texts or represented by substantial finds of grapes, kernels and leaves.

Ceratonia siliqua, Locust-bean or Carob, is grown for its edible sweet legumes. Some consider the tree a native of South Europe. G. Schweinfurth, on the other hand, regarded Arabia as its native home and considered its distribution around the Mediterranean due to ancient cultivation and naturalisation.

Legumes and seeds are found now and then in the tombs, earliest from Kahun, 12th Dyn. Its ancient name was Nutem, and its wood is often referred to in texts from the 6th Dyn. onwards, as a material for making furniture. A bow of carob wood from 1700 B.C. was found at Thebes.

Nuts, like almonds, hazelnuts, walnuts, have been found from Ptolemaic and Roman times. They may have been imported from abroad. They are not recorded from Pharaonic Egypt.

Palms

Three palms are known from Ancient Egypt, the Dom palm, *Hyphaene thebaica*, the Argun palm, *Medemia argun*, and the Date palm, *Phoenix dactylifera*.

Hyphaene still grows wild in Upper Egypt and is a fan palm having the stem branched which is something quite exceptional among palms. Its large glossy fruits have a fibrous tough edible meat of gingerbread taste. Inside is a very hard kernel which has been utilised as a vegetable ivory for making buttons and other objects.

In olden days baboons were trained to gather the fruits. The baboon was a symbol of Toth, the God of knowledge, and hence the Dom palm became Toth's tree and was worshipped, signifying male strength. Its name in olden days was Mama, its fruits Koko. The latter are found in tombs of practically all ages. Its timber of light chocolate colour with black stripes, appearing as dots in transverse section, is hard and durable and was employed for various purposes, among others for making water tubes.

Medemia which is a fan palm like *Hyphaene* but with unbranched stem, was known from Egyptian tombs before the palm was known to science. It has violet fruits of plum-size which were found among offering gifts to the dead and caused much puzzle in the beginning. But their identity was disclosed when Th. Kotschy, the famous Africa-explorer, discovered the palm growing in a Nubian desert valley. The fruits are inedible, but if buried in the ground for some time, the endosperm gets a sweet taste like that of coconut. This practice performed by the Nubians was probably known by the ancient Egyptians and explains why the fruits were given as offering dishes.

Medemia is mentioned in old texts under the name Mama-n-Khanen, and it is also depicted in tombs as a cultivated garden tree.

It was considered extinct in Egypt until a few years ago when it was rediscovered in two places, at Dunkul and at Nakhila near Kurkur, two inhabited oases in the Libyan desert, 220 km southwest of Aswan.

The Date palm, *Phoenix dactylifera*, is unknown in a wild state, nor is it known from which wild species it has sprung. It has always been a human companion and depending on human help for its propagation.

Dates in Ancient Egypt were called Banra or Bonri, the palm itself Benr-t, Benra-t, Bennu etc. There are many name variations. It was a sacred tree symbolising feminity. Sometimes we see the Goddess of the Date palm depicted together with the tree.

57

The date palm has been present in Egypt even before the dawn of history. This is proved by finds of mats made of leaflets of date palm, dating from the Neolithic era. For this reason one would expect that also fruits should be present from this early period. But this is not the case. From the Neolithic time and from the Ancient Empire hardly any date finds are known. From the Middle Empire we know a few finds, but then from the New Empire onwards, there are plenty of dates in practically every tomb. What is the reason?

Maybe the artificial pollination was unknown in early days. Up to the present the Egyptians always propagate their date palms by off-shoots which they separate from the mother tree. Then they are sure to get a male palm from a male and a female palm from a female tree. If sown with seed one could never be sure about the sex until the palm is full-grown and starts flowering. For keeping the date cultivation, seeds are of no importance.

Maybe the Egyptians learnt the secret of artificial pollination from Assyria-Babylonia during the Middle Empire. It is described in detail in the laws of Hammurabi from that period. Or maybe they knew it earlier but did not practice it more generally. In a recent publication I. Wallert says that she has found evidence from ancient texts of its use during the Ancient Empire.

Early Egyptian agriculture

Primitive people always started their development as food collectors, before entering the stage of producing food themselves, or in other words starting some kind of agricultural activity.

The Egyptians were no exception. Before agriculture started along the Nile, they were living on animal hunting and what they could collect of wild plants. Human bodies from that remote period have been found, and Prof. Fritz Netolitzky succeeded identifying the components of their last meal, which consisted of fish, rats, insects, seeds of *Echinochloa colonum*, which still grows as a weed in Egypt, and tubers of *Cyperus esculentus*, still consumed under the name of Habb el-aziz.

How and when agriculture started in Egypt is not yet settled. Certain scholars, e.g. Prof. E. Schiemann, have proposed that it came from Ethiopia, that Egypt thus got it from the south.

This is strongly rejected by Dr. Hans Helbaek. He emphasizes that Ethiopia is a clear secondary gene-centre and could never be the origin of an Egyptian agriculture. There are no ancestral types of wheat and barley in Ethiopia. On the contrary. It is more likely that Ethiopia got its agriculture from Egypt and at a late period (maybe the 1st millennium B.C.?). Although there are no proofs, there is no doubt, according to Dr Helbaek, that West Asia is the origin of Egyptian agriculture, whether it reached Egypt by sea from Anatolia or Syria-Palestine or by land from south Jordan over Sinai, the latter being the most probable.

The oldest cereals found in Egypt are from Faiyum and have been dated to 4,500–4,200 B.C. But evidently this does not represent the oldest agriculture. There must be still older settlements although not yet discovered; see ch. 4a. for epipalaeolithic finds).

In ancient Egypt irrigation was carried out almost exclusively by the basin system. The flood water was conducted to the lands which previously had been divided into basins by earthern walls. Only higher lands were irrigated by lifting the water by means of the 'Shadoof', resembling an ordinary well-sweep with bucket and counter-weight and managed by a single man.

The water having drained off, the land was ploughed as soon as it was sufficiently dry. The seeds were dropped behind the plough as in the 'Talqit' system of nowadays. At the same time the clods of earth were broken by wooden hammers. Flocks of sheep and pigs were driven over the field to tread the seeds into the soil.

The plough of ancient times was exactly of the same type as the baladi plough 'Mihrát', still in use among the peasants. It may be recognised on monuments 4,500 years ago, e.g. in Tih's tomb, Saqqara, as well as in the system of Hieroglyphs.

Cutting was done by means of flint or bronze sickles. Only the ears were cut. The rest of the plants was left to decay in the soil as an organic manure. The ears were heaped and later carried in baskets or nets to the threshing floor.

Threshing was done by driving cattle over the crop, as is still the custom in some parts of Egypt. Likewise winnowing was done by the same kind of wooden forks and fans as nowadays.

The grain was stored in prehistoric times in baskets buried in the sand, as in the Oases at present. Silos of Nile mud and storing houses with holes on the top for filling and doors below for drawing out the grains were used during historical times and partly in prehistoric settlements.

Often the excavated grain show a carbonised appearance. Certain scientists, e.g. R. Biffen, have tried to explain this carbonisation as an action of anaerobic bacteria. This idea is strongly rejected by H. Helbaek. He thinks, that the cereal in ancient days was roasted to facilitate threshing, and when grains were not protected by the hulls, they simply got burnt.

It should be pointed out in this connection that the ancient grains, found in Egyptian tombs, are unable to germinate. When such grain is exposed to the air, the embryo soon gets destroyed. If grain is kept airtight, which is the case in certain peatbogs, they may retain their viability for considerable time. Experiments in Denmark have proved that 1500 years old seeds from Danish peatbogs were able to grow.

Fig. 8. *Triticum monococcum*, probably a mutant of a wild species, was one of the oldest wheats cultivated. From an underground gallery of a pyramid at Saqqara; 5.000 years old. A. Rhachis; B. Spikelets; C. Grain.

CEREALS

Old papyrus documents mention 3 kinds of cereals cultivated in Egypt: JT, BTT or BDT, and SW or SWT. How these words should be pronounced, nobody knows, but to judge from the corresponding Coptic words most probably JATE, BOTET and SWO or SWOT.

Jate was barley, *Hordeum vulgare*, of the same slender type as is still cultivated in Egypt under the name of Manshuriya-barley. In the oldest known Neolithic settlements, the Faiyum culture, Hans Helbaek (unpublished work) found 4 different species of barley, *Hordeum distichum*, *H. deficiens*, *H. hexastichum* and *H. vulgare*. But in later Neolithic tombs and throughout the Pharaonic era, only *H. vulgare* is present. Also germinated barley is frequently found. It played an important role in the Osiris rites as a symbol of resurrection.

Botet was emmer-wheat, *Triticum monococcum* or *Tr. dicoccum*, which has not been cultivated in Egypt during the last 2,000 years. It differs from modern wheats in having a brittle axis, its glumes tightly enveloping the grain thus making the threshing difficult. Nowadays emmer is replaced by socalled naked-wheats, having a tough axis and dropping its grain easily.

The identification of Swo or Swot is still an unsolved problem. It may have been a type of modern wheats, in such a case probably of *Triticum durum* strain. But there is also a possibility that it did not represent a grass cereal at all but another bread plant, *Ensete edule*, of which the inner part of the stem still constitutes an important starch food in Ethiopia. No substantial *Ensete* find is known from ancient Egypt, but the writer has interpreted certain plants painted on prehistoric pottery from the Gerzean (Nagada II) period as being *Ensete*.

Breads of barley and emmer are frequently found, and it is evident that yeast may have been used for fermentation. A yeast fungus, *Saccharomyces winlocki*, has been identified from the tombs. Sometimes lichens were imported from Greece and Anatolia and added to the bread to make it more porous and tasty. Such lichens are still imported from these countries under the name of Sheba and used for bread-making up to the present.

Beer in ancient Egypt was made of half-baked bread. This was broken into pieces, water poured over it and the liquid left to ferment. The milk-white alcoholic Boza of the Nubians in Egypt is still made in this way.

WEEDS

Very often the stored grain found in the tombs is mixed with weed seeds of different kinds. Due to this we have got a very good concept about the weed flora of ancient Egypt, which seems to have been about the same as nowadays. Altogether 50 weeds are represented by substantial finds, and the richest represented families among them are *Compositae, Gramineae, Leguminosae* and *Cruciferae*. The same families still dominate the Egyptian weed flora.

An interesting weed is the leafless parasite Dodder or *Cuscuta*. Its seeds germinate in the soil, but as soon as the stem develops it searches for a suitable host, and develops haustoria which penetrate the tissues of the host and sucks food from it. Once it is well established it loses connection with the soil and continues living as an aerial parasite. One of these dodders, *Cuscuta pedicellata*, has been recognised among the ancient weeds.

Another interesting weed is Darnel, *Lolium temulentum*. It is a grass which is poisonous due to a fungus parasitizing in the grains and pro-ducing the deadly poison temuline. The same fungus attacked the grains in ancient Egypt, and has been found in 3,000 years old grains.

On the whole, the ancient Egyptian fields were remarkably clear of weeds as compared with grain fields of other countries. This is a fact, which has also been emphasized by several classical authors, especially from the Graeco-Roman time.

PULSES

Pulses have always constituted the main source of protein for the Egyptian people, who as a rule cannot afford to eat meat.

The most popular pulse in ancient as in modern Egypt is a small-seeded strain of Broad-bean, *Vicia faba*, called in Arabic Foull. It is eaten every morning by almost everybody.

Other pulses which have been found in the tombs and which are still in cultivation, are: *Cicer arietinum*, Chickpea or Hommos; *Lens culinaris*, Lentil or Ads; *Trigonella foenum-graecum*, Fenugreek or Helba; *Lupinus albus*, Lupine or Tirmis; *Vigna sinensis*, Cowpea or Lubia.

Lathyrus sativus, Grass-pea or Gilban, and *Trifolium alexandrinum*, Egyptian clover or Berseem, were cultivated for fodder in ancient as in modern time.

Pisum sativum, Pea or Besilla, is probably a modern introduction. A record of it from Hawara, Graeco-Roman time needs verification.

Oil and Textile plants

Vegetable oil was much in demand in ancient days, even more than nowadays. Alcohol distillation was not yet invented, and the manufac-turing of perfumes and unguents required fats and oils for fixing the essences. Also lamp oils were of vegetable origin, the most common ones being Castor oil of *Ricinus communis*, and Olive oil of *Olea europaea*.

Olive stones have been found in the tombs from the 18th Dyn. onwards. Probably the olive was introduced into Egypt from Syria and Palestine during that period. It may look strange that only kernels and not the entire fruits were given to the dead. But in olden days it was considered enough with a symbol of the fruit like the kernel. Osiris could afterwards transform these symbols into real food.

Besides the fruits, one frequently finds olive leaves and branches as funeral decorations. The famous 'crown of justification', which was put on the forehead of the mummy as a sign that he was free from sins, was made of olive leaves.

An interesting find was made by the University of Michigan expedition in Kom Aushim-Karanis near Faiyum. They found 2,000 year old fodder cakes made of remains after olive pressing. These cakes are exhibited in the Agricultural Museum in Cairo.

Other appreciated vegetable oils in ancient days were obtained from *Lactuca sativa*, Lettuce; *Linum usitatissimum*, Flax; *Carthamus tinctorius*, Safflower; *Raphanus sativus*, Radish; *Sesamum indicum*, Sesam. The oil from *Balanites* and *Moringa* has already been discussed. Almond oil from *Amygdalus communis* is also recorded, apparently it was pressed from imported nuts.

The main textile fibre during the Pharaonic era was Flax. Already during Predynastic times people could produce the most exquisite linen. Cotton is of late introduction in Egypt. The oldest known find is from the 5th cent. A.D. from the Coptic monastery of Phoebammon near Thebes. The earliest mention of cotton in European history was by Alexander's soldiers, who on their return from India reported that they had seen shrubs carrying wool.

Ropes and baskets were made in the same way as now. Egyptian baskets have a filling of halfa grasses, *Imperata* or *Desmostachya*, outside wrapped in leaflets of datepalm. Modern and ancient baskets are identical, so also the ropes.

An exception is a type of basket coffins, which preceded wooden coffins for the mummies. They are all made of one and the same plant, *Ceruana pratensis*, a strong-smelling composite, still growing wild along the Nile Khores of Upper Egypt.

Ceruana is the Heden-plant of ancient Egypt, of which brooms were made for sweeping off evil spirits from the temple floors. It is evident, that the coffin made of it also should protect the mummy from being attacked by evil spirits. *Ceruana* is still used for making brooms, although the meaning behind the choice of this plant is nowadays forgotten.

Vegetables

Three types of onion were cultivated in ancient Egypt, *Allium sativum*, Garlic or Thoam; *A. cepa*, Onion or Bassal; and *A. kurrat*, Salad-leek or Kurrat. The last is hardly known outside Egypt and is always consumed raw. The onion was ascribed magic power, and people used to call its name when swearing their oaths. This made Juvenal and Pliny joke

63

about the people who let their Gods grow in their vegetable garden. Egyptian peasants still suspend a bundle of onions above their door or smear onion juice on their threshold to protect the house from evil spirits.

Onions were also used in mummification. Thus onions were placed on mummies to stimulate the defunct to breath. Ramses III had small onions inserted into his orbits to serve as artificial eyes.

Melons and cucumbers of all kinds have always been grown in Egypt, especially on sandy Nile banks, which yield the best crop. Up to the present there is an abundance of endemic varieties cultivated in Egypt. Certain of these have been found in the tombs and are of very ancient cultivation. It is said in the Bible, that the children of Israel during the Exodus longed to go back to the country which had given them such delicious melons and cucumbers.

It is custom in Egypt also to chew melon seeds, 'Libb', which are roasted, and salted. The best libb is of brown colour and derives from a small-fruited bitter variety of water-melon, *Citrullus vulgaris* v. *colocynthoides*, which is cultivated exclusively for its seed in Upper Egypt and the Sudan.

Such brown libb has been found in the tombs, earliest from the Temple of Sahure, 5th Dyn., the latest in tombs at Gebelen from Graeco-Roman time. The habit of chewing libb is thus a very ancient tradition.

Among the most popular vegetables, now like in ancient times, is the long white Radish, *Raphanus sativus* v. *aegyptiacus*, of which both root and leaves are eaten raw, and the long-leaved lettuce, *Lactuca sativa* v. *longifolia*, depicted in almost every tomb as a precious gift on the offering tables. The latter was also a symbol of fertility, and the Fertility God, Min in the Old and Amon in the New Empire, is often seen holding a lettuce plant in his hand.

Celery, *Apium graveolens*, was well-known as a vegetable, but at the same time as a symbol of sorrow, not only in Egypt but in ancient Greece as well. It was planted on tombs and used in funeral decorations. Such a celery garland, found around the neck of a mummy is exhibited in the Agricultural Museum in Cairo.

A vegetable rarely seen outside Egypt, is the violet carrot with white centre, *Daucus carota* v. *boissieri*. It must be of very ancient cultivation, although not yet found in the tombs.

The same is the case with *Corchorus olitorius*, which constitutes the popular Egyptian dish 'Melokhia'. It has not yet been found in the tombs but may be expected.

Cabbage, *Brassica oleracea*, is known from the Graeco-Roman time onwards. Spices like Dill Fennel, Coriander, Anis, Cumin, are all well-known from ancient Egypt. Also *Beta vulgaris* and *Malva parviflora*, of which the leaves are still eaten as a pot-herb, are known by substantial finds. These two also grow wild besides being cultivated.

Garden flowers

Horticulture was highly developed in ancient days. Wealthy people always had a garden in front of their house with shady trees, pergolas of climbing vine and usually a pond with papyrus and lotus. The present habit of putting flower pots along the paths of the garden seems to be of Pharaonic origin, at least to judge from certain tomb representations.

Wild flora was used in abundance in their funeral garlands. The ancient garlands were narrow and elegant so as to fit inside the coffin. On a string of date-palm fibre or other material were folded leaves of the sacred Shoab-tree, to serve as holders for other flowers, usually wild ones, which were sewn together with these leaves. Due to these wild flowers we know when the funeral took place, if in winter or summer, if in spring or autumn. As a yellow ornament were chosen flower-heads of *Acacia seyal* and *nilotica*, *Picris radicata*, *Chrysanthemum coronarium*, *Senecio desfontainei*, *Helichrysum stoechas*, *Sesbania sesban*, *Carthamus tinctorius*, *Ambrosia maritima*, etc. A blue or mauve effect was obtained by flowers of *Delphinium orientale*, *Centaurea depressa*, *Anemone coronaria*, petals of *Nymphaea coerulea* and others. For red and rose colours were chosen *Papaver rhoeas*, *Punica granatum*. *Epilobium hirsutum*, *Alcea ficifolia*, *Lychnis coeli-rosa*, *Matthiola livida* and also the red berries of *Withania somnifera*. For a silvery effect *Celosia argentea* was used.

We also find in these garlands artificial flowers made of pith of *Scirpus*- and *Cyperus*-species.

Among ornamental flowers, the fragrant ones were the most popular, as still is the case in Egypt. We find in the tombs branches of *Myrtus communis*, *Origanum majorana*, *Rosmarinus officinalis*, *Mentha piperita*, *Laurus nobilis*, *Jasminum sambac* and *grandiflorum*, *Narcissus tazetta*, *Lawsonia inermis* etc. etc.

Lawsonia inermis, the Henna shrub, was appreciated not only for its fragrant flowers, 'Tamr henna', but also for its leaves which were dried and powdered and if stirred with water yielding an orange-red dye. Egyptian ladies have used this dye throughout history for their palms, toe- and finger-nails, hair, etc. The native home of henna is unknown, but it may have been introduced from Iran, as henna is a Persian name. G. Schweinfurth, on the other hand, believed that it was introduced from Arabia with the Punt-expeditions.

The sacred Shoab-tree, which we see in practically every tomb as garlands or bouquets, has its own history. It is told in ancient texts, that the Gods when entering the Nile valley from the south, brought with them a sacred tree, Shoab, which was planted outside the temples. This tree changed name during the Graeco-Roman period, when it was known as Persea. During the early Islamic period it changed name again and was now known as Lebbakh. But soon after, all knowledge about it was forgotten.

domestication was later and played only a subordinate role in animal husbandry. The wild pig, possibly *Sus scrofa barbarus* Sclater. palaearctic in origin, became apparently extinct by the end of the 19th century.

Ass

Probably derived from the wild ass now (almost?) extinct, played a very important role in ancient Egyptian economy.

Horse

The Hyksos invasion (about 1675 B.C.) brought the horse which appears then frequently in wall decorations and on furniture; it was a robust breed and may have come from the steppes of the Caspian.

Dog

Two races are present, a greyhound type with standing ears and short coat, probably derived from wild Canidae of North Africa; similar dogs are frequent in rural districts of the Sudan today. A race with short legs appears later than the 'greyhound' type.

Cat

A probably wild but tamed species appears already in Neolithic sites; domestication from *Felis silvestris lybica* Forster is apparent during the New Kingdom (1550 B.C. onwards); it became a cult animal with the 'Bastet' cult flourishing at Bubastis and many mummified specimens are present.

Camel

Few representations exist of the camel on war scenes from the 19th dynasty; it was apparently not domesticated in Egypt.

Non-domesticated wild animals appear frequently in pictorial representations; some came from the surrounding desert such as gazelles kept as pets (*G. dorcas dorcas* and *G. dorcas isabella*, *G. soemmeringi*, *G. leptoceros*, *G. dana ruficollis*, *Ammotragus lervia*, *Dama mesopotamica*) *Oryx* and *Addax* seem to have been semi-domesticated and appear as sacrificial animals.

Frequent are the scenes depicting the arrival of imported animals: *Papio hamadryas*, *P. anubis*, *Erythrocebus pyrrhonotus* from 'Nubia' or the 'land of Punt'; the same source supplied the geppard *Acinonyx jubatus*.

70

Some scenes seem to suggest a semi-domestication of the ostrich; otherwise only *Anser anser* the goose seems to have been fully domesticated. Wild fowling in the still existing swamps was a favourite pastime of the ancient Egyptians judging from the numerous pictorial representations. The following species are mentioned by Boessneck: *Anas acuta, A. platyrhyncha, A. crecca, Casarca ferruginea, Tadorna tadorna, Anser albifrons, Oxyura leucocephala, Spatula clypeata, Nyroca fuligula, Cygnus olor, C. cygnus, Pelecanus onocrotalus, Fulica atra, Grus grus, Anthropoides virgo, Ardea cinerea, Thresciornis aethiopicus, Streptopelis turtur, Columba livia.* The presence of extensive marshes is not only depicted but is clearly shown by the number of aquatic birds now only seen on the Upper White Nile. This richness was in time severely depleted by continuous hunting. A further testimony for the birds of ancient Egypt comes from ch. III of Meinertzhagen (1930) (R. E. Moreau. Birds of Ancient Egypt). Besides pictorial representations, information comes from mummified specimens and hieroglyphic signs. Pictorials are best up to the IV dynasty, later standards decline; there is never any veracity as to size and colours are crude (see ch. 6b).

The pictorials of Ancient Egypt also contain fishes and these have been described in several papers e.g. Gaillard (1923) and Keimer (1948). Fishponds were kept with *Tilapia* faithfully painted.

REFERENCES

Boessneck, J. 1953. Die Haustiere in Altägypten. Veröfftl. Zool. Staatssammlung München 3: 1–50.

Gaillard, C. 1923. Recherches sur les poissons representés dans quelques tombeaux égyptiens de l'ancien empire. Mem. Inst. Français Archéol. Orient. Caire.

Keimer, L. 1948. Quelques representations rares de poissons égyptiens remontant a l'époque pharaonique. Bull. Inst. Égypte 29: 263–274.

Moreau, R. E. 1930. The birds of ancient Egypt. In: Meinertzhagen Nicoll's Birds of Egypt vol. 1.

Zeuner, F. E. 1963. A History of Domesticated Animals. Harper & Row, New York.

4d. RIVER AND NILE VALLEY BEFORE MANS' INTERFERENCE

by

J. RZÓSKA

In previous chapters the changes in the environment of the Nile valley during prehistoric and early historic times have been traced. In Egypt during dynastic times there had already been a great alteration in the natural scene. Man's influence was decisive in keeping riverain Egypt green by irrigation; further south his efforts at that time were negligible.

Interest in the life-giving river changed from ancient Egyptian hearsay, worship and fear (when the flood failed) with the arrival of the Greeks and Romans. Herodotus (about 460 BC) recorded the tales of his Egyptian informers often with incredulity; he travelled to the II cataract to find out where the water came from, doubt turned into inquiry.

The practical Romans, when in firm possession of Egypt (100 BC to 400 AD), began to gather intelligence about the country beyond their domain. Judging from the many Roman writers who commented on the geography, the natural history and, above all, the role of the Nile, the interest was intense (Ptolemy's map – 150 AD). It may have been a Roman who coined the words 'Aut Nilus aut nihil' which in lapidar form sums up what we know now. Having pushed their military forays deep into Nubia in 23 BC, their curiosity was still more stimulated. The military reconnaissance ordered by Nero (54–68 AD) penetrated deep into the Nile valley before being stopped by 'impenetrable' swamps; it is not certain if these were the swamps on the White or Blue Nile. Kirwan (1957) relates that at Meroe, between the Astaboras (Atbara river) and the VI cataract, the emissaries reported finding 'parakets, monkeys, tracks of rhinoceros and elephant' – obviously a savanna environment. During Meroitic times, from about 725 BC to 350 AD, temples and other monuments were built, e.g. at Naga and Musawwarat es Safra, which lie now in the desert but must have been originally in a more favourable environment. There is an engraving in the Naga temple showing a Meroitic king riding an elephant. Christian civilisations flourished along the Nubian Nile valley from the 6th to the 14th or 15th century, but their magnificent murals are austere and inward looking. Arkell (1961) has given a concise and lucid history of that part and the rest of the Sudan; the Christian monuments of Faras and Dongola have been recently exposed as testimonies of a thriving culture close to the river.

The Arab conquest of Egypt in 600 AD, and later the penetration into Nubia, brought chroniclers into the country, amongst them Idrisi in the

12th century whose description of the country has survived (Geographie d'Idrisi, translated from Arabic by P. A. Jaubert, Paris 1936); it was unavailable to the present writer. Idrisi and probably other Arab travellers have commented on the nature of the country. The advance of the Arabs into the central Sudan in the following centuries was finally halted by the biological environment, Tabanids carrying diseases killed their horses and camels (see ch. 23).

Few European explorers penetrated south before the 18th and 19th centuries. It is then that the great upsurge of interest in Egypt and the Nile valley became widespread. In 1789 Napoleon invaded Egypt, then under distant Turkish rule but semi-independent; he brought with him a large number of 'savants' who explored the country and left us the 'Description d'Égypte' in 24 volumes (Paris 1820), an outstanding achievement. Shortly before, a Frenchman, Sonnini de Manoncourt, observed the country, probably as a royalist intelligence officer. He gave an interesting description (1799) and commented on the fauna; he related the fate of the last hippos killed in the Delta at the beginning of the 17th century.

The 19th century saw an ever increasing spate of travellers, explorers and adventurers, first mainly attracted by the antiquities and in the second half by the 'quest for the sources of the Nile'. Sir H. Johnston (1906) has written an impressive 'Record of the exploration of the Nile and its basin'. More recently, A. Moorehead has narrated an eloquent story of the White and the Blue Nile (1960, 1964). Before them, 'compendia' appeared collating useful information for travellers (Johnston 1878 and Reclus translated into English about 1885–6). From these and other contemporary sources the state of the Nile valley with an as yet untamed river is briefly reported here.

Great inundations occurred in many places; lake-like conditions existed north of the Aswan cataract (Burckhardt 1822), leaving temporary lakes and pools, causing the emergence of insect pests and diseases. On the White Nile above Khartoum all travellers comment on the wide expanse of the river up to '2 miles' broad with swampy inlets and widespread inundations. In the Sennar region of the Blue Nile Bruce, who travelled here and in Abyssinia from 1768 to 1773, comments on the 'lakes' formed by the river floods and the complete change of the landscape during the succeeding dry season. Baker (1871) was probably one of the first to describe the astonishing change of the Atbara in flood and reduced to pools in the dry season.

In many descriptions of the riverain scene in Upper Egypt and Nubia, observed mostly by travellers from boats, the narrowness of the alluvial cultivation and vegetation is mentioned, interrupted by completely rocky and barren stretches. In 1850 Bromfield saw no reeds and marsh plants there, only in some few places flat inundations allowed few groves of *Tamarix* and *Acacia* and date palms to develop 'pleasingly'. Otherwise only irrigated strips were green and this has not changed to date. But up the White Nile all observers from 1840 onwards were impressed by the luxuriant development of fringing reeds and water vegetation. 'Archipelagos' of islands were covered with luxurious forests festooned with climbers between Dueim and Abba island and evoked admiration. But some careful observers noted already ominous changes. Heuglin who came here twice, in 1852 and 1863, saw the rapid depletion of *Acacia* forests between Geteina and Dueim; the great Schweinfurth commented on the further decline in 1868. By the time Muriel wrote his report on the Sudan forests (1901) the damage was ruinous. Ship builders and fuel providers were the culprits unchecked by any authority. The floating islets of the White Nile, composed of *Vossia*, *Papyrus*, *Pistia* and Ambatch, were observed in their northern travel; when stranded they were colonised by secondary vegetation, consolidated by their roots and narrowed down the river channel in some places. This phenomenon increased southwards and found its culmination in the 'Sudd'.

As far as human records exist the Upper Nile swamps have been an obstacle to traffic. All rivers and many of the standing waters were and are fringed with vegetation (see ch. 12) which constantly invaded the free water either by expansion or by breaking off the shore through the combined forces of wind and current. These floats joined through the entanglement of their roots and became blockages. Every traveller in the 19th century had a bitter tale to report. In 1864 the whole breadth of the White Nile near the Sobat was blocked (Heuglin); in 1880 the river traffic between Khartoum and Lado was closed for almost a year. On the Bahr el Ghazal 'boundless seas of grasses' occupied the upstream part near Meshra er Req. The Turkish Governor-General of the Sudan at that time, Ismael Eyoub Pasha, employed a special engineer, E. Marno, to keep the Bahr el Gebel free for administrative/political reasons. Marno's work, undertaken with a great labour force, had no permanent effect. It was Marno who finally rescued the ill-fated Romolo Gessi, Governor of the Bahr el Ghazal province. Gessi was returning to Khartoum from Meshra er Req in a steamer with a number of tow boats under enormous difficulties, when finally on 25th September 1880 they were enveloped by a vegetation block several thousand meters long. All attempts to free themselves failed, disease and famine broke out and ended in cannibalism. At literally the last hour they were reached by Marno on 4th January 1881.

From 1882 to 1898 the whole of the Sudan was under the Mahdi's Islamic revolt against the Turkish-Egyptian regime. When the country was re-occupied by the Anglo-Egyptian Government in 1898 all swamp rivers were found blocked up. According to R. Hill, historian (and regional manager of steamer traffic), 'it took until 1905 to clear a channel through the Bahr el Gebel'; the Ghazal was made navigable to the provincial capital, Wau, only by 1904.

There must be an error in the book on Alexine Tinne's travels written by Penelope Gladstone (1970), her relative and chronicler. Relating the difficulties of travel which this ill-fated expedition had to endure in the Bahr el Ghazal, she speaks of the lake-like expanse (Ambadi) where the steamer 'paddles were constantly clogged with water hyacinth and other plants' (p. 129). This was in 1863; a few years later the careful botanist, Schweinfurth, mentions mainly *Vallisneria* but not the conspicuous *Eichhornia*. During the visits of the Hydrobiological Research Unit of the University of Khartoum (1953–1956) the composition of the luxurious bottom vegetation was again quite different. The great water hyacinth invasion occurred only in 1957/8 (ch. 21).

The advance of desert conditions in the north was observed by Hoskins who travelled in Nubia in 1833. Comparing his impressions gained at Meroe with those of Cailliaud of 1819, he concludes that the desert had advanced even during the last 14 years; traces of savanna apparently were still able to support lions and their prey.

Almost every traveller has commented on the fauna and to compile this abundant information would require more space than justified in this book, but it would provide the documentation for the recent withdrawal of the large mammals first due to desiccation and later to man.

Observations of specifically limnological interest are few in the early accounts. The colour of river water and its changes was known to the ancient Egyptians who apparently painted the Nile god, Hapi, green and dark according to brown silt laden flood and the green colour of the low river. Petherick (1869) notes the change in Nubia where the green colour in August changed to brown in a fortnight. He also remarked on the 'pond'-like appearance of water at Gebelein on the White Nile. Baker described the water of the White Nile near Abba Island as 'like an English horsepond'. Gleichen wonders about the green appearance of the White Nile at Dueim in May and speculates on its origin; these may have been water blooms. Some of the first observations on currents were made by Peel at the junction of the two Niles in 1851; Baker measured currents and discharges during his journey up the Bahr el Gebel and also gave the first assessment of area and depth of Lake No.

A tremendous amount of information on the state of the Nile country is contained in all the accounts of these explorers and scientists; but almost every one of them has also reported on the increasing desolation and depopulation, especially in the south. A sinister type of adventurer began to penetrate, slave traders and ivory hunters. In the fifties of the last century under a corrupt and inefficient government, large expeditions went upstream organised by traders in Khartoum and run by ruthless middlemen. The extent of the slave trade has been dramatically described by A. Moorehead in his monograph of the White Nile. In the Bahr el Ghazal whole tribes were decimated and many years later remnants of languages once spoken were collected by dedicated people like the Verona Fathers.

The ivory hunters, often slave traders as well, found their paradise to despoil. Casati (1891) who was in the entourage of Romolo Gessi, Governor of the Ghazal Province, from 1853 to 1879 reports that about 4 million kg of ivory were exported during that time to Khartoum. Even Europeans joined in this lucrative trade; one of them was J. A. Vayssiere who left his memoirs and notebooks, now in the Société de Géographie de Paris (MS 8° 23). Near Shambe he saw nearly 1,000 elephants (I am indebted to R. Hill for making available a typescript for perusal).

When the country was re-opened by the Anglo-Egyptian reconquest, one of the first tasks undertaken was a survey of the whole country by army personnel and later by medical and agricultural teams. Gleichen has compiled two volumes of a 'compendium prepared by officers of the Sudan Government'. This remarkable work includes a comprehensive description of the state of this vast country including almost every aspect and even the vegetation, history, archaeology, the state of the river, people and fauna. From there onwards the peaceful development of the country to the present independent Republic had started.

REFERENCES

Baker, Sir Samuel White, 1866. The Albert Nyanza, great basin of the Nile and the exploration of the Nile sources. MacMillan, London 1866.
Bromfield, W. A. 1856. Letters from Egypt and Syria. Printed for private circulation only. London 1856.
Bruce, J. 1790. Travels to discover the source of the Nile in the years 1768 to 1773. Edinburgh.
Burckhardt, J. 1822. Travels in Nubia by the late John Lewis Burckhardt. London 2 ed.
Cailliaud, Fr., 1826. Voyage à Meroe au fleuve Blanc, au-delà de Fazogl dans le midi du Royaume de Sennar, à Syouah et dans cinq autres oasis 1819–1822, 4 vols. Imprimerie Royale Paris 1826.
Casati, G. (Major) 1891. Ten years in Equatoria and the return with Emin Pasha. 2 vols. transl. from the original Italian by the Hon. Mrs. J. Randolph-Clay. London and New York, F. Warne & Co.
Gladstone, Penelope. 1970. Travels of Alexine Tinne (1835–1869). J. Murray, London.
Gleichen, Count. 1905. The Anglo-Egyptian Sudan, vol. I & II. A compendium prepared by officers of the Sudan Government. London H.M. Stationary Office.
Heuglin, Th. von. 1869. Reise in das Gebiet des Weissen Nil und seiner westlichen Zuflüsse in den Jahren 1857–1862. Leipzig (Map redrawn-Justus Perthes Gotha).
Hoskins, G. A. 1835. Travels in Ethiopia, above the II cataract, exhibiting the state of that country and illustrating the antiquities, Arts and History of the ancient kingdom of Merowe. London.

Johnston, Sir Harry. 1906. The Nile Quest. A record of the Exploration of the Nile and its Basin. Alston Rivers, London.

Johnston, Keith ed. 1878. Stanford's Compendium of Geography and Travel based on 'Hellwald-Die Erde und ihre Völker'. E. Stanford, London.

Irby, Ch. & Manglis, James 1844. Travels in Egypt and Nubia, Syria and the Holy Land etc. J. Murray, London.

Junker, W. J. 1890–92. Travels in Africa during the years 1882–1886. Transl. by A. Keane. Chapman, London.

Kirwan, L. P. 1957. Rome beyond the southern Egyptian frontier. Geogr. J. 123: 13–19.

Moorehead, A. 1960. The White Nile. Penguin ed., 1963.

Moorehead, A. 1962. The Blue Nile. Four Square edition, 1964.

Muriel, C. E. 1901. Report on the forest of the Sudan. Al Mokattam Press, Cairo.

Peel, W. Captain R. N. 1852. A ride through the Nubian desert. Longman & Co., London.

Petherick, J. 1861. Egypt, the Sudan and Central Africa, with explorations from Khartoum to the regions of the Equator. Blackwood, Edinburgh and London.

Reclus, E. 1885–6?. The earth and its inhabitants. Universal geography. ed. by A. H. Keane. Vol. X. J. S. Virtue & Co., London.

Russeger, J. 1843, 1844. Reisen in Europa, Asien und Afrika, unternommen in den Jahren 1835 bis 1841. II Bd. Aegypten, Nubien und Ost-Sudan. Schweizerbart, Stuttgart.

Schweinfurth, G. A. 1873. The Heart of Africa, travels and adventures from 1868–1871. Transl. Sampson Low, London.

Sonnini de Manoncourt, C. S. 1799–1800. Travels in Upper and Lower Egypt, undertaken by order of the old government of France. Transl. J. Debrett, London.

Walker, F. A. 1884. Nine hundred miles up the Nile, Nov. 3d-Feb. 9th 1884. West & Neumayer Co., London.

Werne, F. 1848. Expedition zur Entdeckung des Weissen Nil (1840–1841). G. Reimmer, Berlin.

5. VEGETATION OF THE NILE BASIN

by

J. RZÓSKA & G. E. WICKENS

No attempt is made here to give a complete account of the flora; excellent books on this subject exist. Vegetation is used here only as indicator of environmental conditions.

Two primary factors govern the present extent of the vegetation along the Nile, the climate and the hydrographic network. The map of the drainage basin (see Fig. 14) resembles an enormous plant, its water supplied by its roots inside Africa, with a lanky stem piercing the arid and finally waterless north and ending in the 'crown' of Egypt. The great changes in the past in Egypt and Nubia resulting in the present condition deserve greater attention than the southern part of the Nile basin.

In a general paper on rainfall and vegetation in north-east Africa Kassas (1955) has described three major vegetation zones for Egypt and Nubia: the northern zone of mediterranean (winter) rains, the central rainless zone from Quena to the III cataract and the southern zone of summer rains which starts at Karima with 25 mm of rainfall and gradually increases southwards (Fig. 10).

The climate of Egypt has clear cut features; the mediterranean winter rains, from November till March, are small and have considerable amplitudes (see Fig. 5). Alexandria has 184 mm with a variation of 29%, Cairo only 38 mm with 65% variation. Further south rain is exceptional. Following this rain pattern vegetation is widespread along the coast but narrows down to the Nile valley in Upper Egypt. The desert represents a special environment.

A third factor influencing the vegetation is man, who brought profound changes especially in cultivated areas. V. Täckholm has supplied some remarks additional to her chapter (4c.). Seven regions are recognised in Egypt:

1. The Nile region, comprising the Delta, the river valley and Faiyum fed by the Nile through the Bahr- el Yussuf canal.

2. The desert region, subdivided into the Libyan desert west of the Nile, and the eastern deserts extending to Sinai. Southwards to and into the Sudan a northern and southern part can be distinguished. The Egyptian deserts belong to the Sahara-Sindian phytogeographic region, which stretches across northern Africa, Arabia into India. Some isolated mountains within have Irano-Turanian floristic elements.

3. The oases of the Libyan desert.

4. The mediterranean region, floristically part of that basin's province.

5. The Red Sea coast.

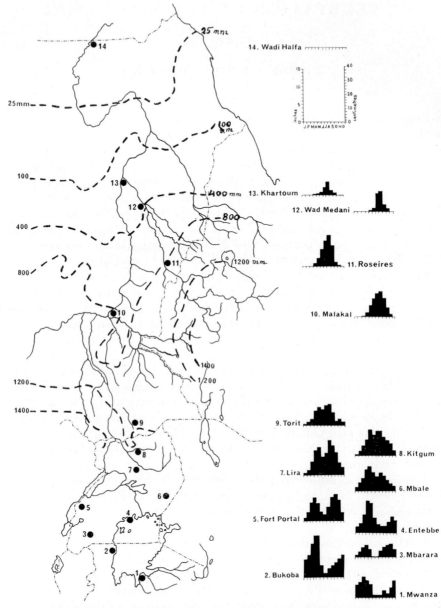

Fig. 10. Rainfall along the Nile system (except Egypt, which is rainless with small Mediterranean belt, see Fig. 5). Main isohyets and diagrams of seasonal distribution of rain; note narrowing down of seasons towards the north; from various sources.

6. Gebel Elba in the south eastern corner of Egypt, its flora akin to the Sudanese Red Sea Hills.

7. Sinai; the southern mountain region is Irano-Turanian.

The flora of cultivated lands in Egypt has recently been studied by several botanists, especially by J. Kosinova; along the water courses it consists mainly of synanthropophytes; habitats are mainly man-made, though they exist also semi-natural places, only temporarily influenced by man. This 'weed' flora may be everywhere or specifically riverine or outside the river's influence on the sea coast. Many of these plants are known from Neolithic times onwards (see Ch. 4c) others have been introduced with rice culture. Of alien plants *Eichhornia* is a special case; introduced about 50 years ago by amateur gardeners and now present in some canals and Lake Edku, it is a potential menace and kept under surveillance.

Many plant species have disappeared due to the change from basin- to perennial-irrigation, e.g. the genera *Alisma* and *Damassonium*. The swamps of Lower Egypt have been under cultivation for many centuries and many marsh species have disappeared. The two heraldic plants of ancient times, papyrus and lotus were seen last in 1820/21 near Damietta and in Lake Menzalah. The discovery of a stand of 20 papyrus plants was therefore a great surprise (Hadidi 1971); they were found in a marsh near Umm Risha in the Wadi Natrum chain of depressions, which form such a conspicuous feature even in a space photograph (fig. 17b). *Nymphaea lotus* var. *aegyptiaca* was found in a small drainage channel near Beni Suef, opposite the Faiyum (Hadidi 1971). Two zoologists, Meinertzhagen (1930) and Moreau (1966), have commented on the broad aspects of the vegetation in Egypt and its divisions: the Delta an 'oasis', the mediterranean strip along the coast with alkaline flats and some swamps, the contrast between the cultivated land and the ever menacing desert. The Faiyum, though much changed during its long history, still bears testimony to some ancient features of the Nile valley. On the whole the flora of Egypt is poor, Meinertzhagen counts '1514' species. Kassas has described a.o. the Delta coast (1955) with its lakes (see ch. 20), separating the cultivated land from the Mediterranean and salt marshes with halophytic vegetation surrounding the lakes. Here the shoreline is receding because irrigation dams have diverted the deposition of Nile sediments for many centuries. Sand dunes extend east and west of Lake Borullus, some moving inland. The richest vegetation is found west of Alexandria comprising some 800 species, 'half of the Egyptian Flora'. Owing to the regular winter rains here a large number of ephemeral species appear every year. South of Cairo the vegetation shows the sharp division between the Nile valley and the desert. Hadidi describes the flora of the Nile valley in a following contribution; the impact of the new inland lake of Aswan provides prospects of great interest.

Moreau regards most of Upper Egypt as 'Saharan'; now even the river valley in a large part has been drowned and sand dunes descend to the river's edge. The desert itself is an extreme environment. Bagnold travelled through the pebbly western part and saw 'two scraps' of living

81

vegetation during a route of 270 miles (430 km). Newbould & Shaw (1928) found only a few 'age old tufts of nissa grass' (*Aristida* sp.). Yet the desert has a life of its own and there is an 'intimate relation between the landform and the vegetation' (Kassas 1953a). In the Egyptian part of the desert Kassas distinguishes a number of surfaces, all exposed to lack of water, high insolation and deflation by winds, but even minute differences evoke vegetational responses. Fissures in rocks, drainage furrows and partial shade are such factors; besides, inaccessibility on cliffs has created some refuges for desert plants.

Desert wadis are a special habitat; Kassas (1953b) has examined some in the eastern desert near Helwan. Two types of plants exist, ephemeral species which appear annually depending on rain, and perennial species; various micro- habitats harbour different floras. Crevices in rock have 26 species, shallow accumulations of alluvial wadi soil are richer with 51 perennials and about 60 ephemerals, deeper soils bear 49 and patches of grass land 34 species but with a denser cover. The climax of this flora is on wadi terraces with a dense growth of scrub, often severely depredated by man. Some of these wadis are drainage lines with a considerable catchment area and offer better conditions to plant life than the surrounds, and the average rainfall records. Meinertzhagen (1930) counts 193 plant species peculiar to the desert. The large depressions in the western desert, where the ground water level is in places within reach of some plants, are regarded by Moreau (1966) as 'ecological islands'.

Before the Wadi Halfa area was drowned by the Nubian part of the Aswan basin, a group from the University of Khartoum reported on its biology (Pettet *et al.* 1964). From 1902–1960 rainfall recorded was at 3 mm per year; a freak rain of 19.4 mm in July 1950 was remembered by the local population. The range of temperatures is enormous from 52.5° C in April to a minimum of —2° C in December 1917. The daily changes are also great; at the time of the visit they ranged from 42 at noon to 22° C at night. Relative humidity was 20% but may fall to zero in some micro-climatic situations; on Nile silt a surface temperature of 84° C was recorded. The existing fauna, poor in species, has to adopt a clear diurnal rhythm of activities, seeking shelter during the day. The flora consisted of 54 species of flowering plants mainly agricultural weeds including 4 species of *Acacia*. The once famous date palms of this region are now drowned. They survive further south in the part which is not influenced by the new water basin of Lake Nubia (Aswan basin).

Kassas (1956) investigated the desert flora at Omdurman, 1500 km south of his Helwan sites. In spite of a summer rainfall of 163 mm according to long term statistics, conditions are severe. The rain is unreliable, it may fall only during 4 months with 8 months rainless. Sharp fluctuations from year to year may be expected (Kassas 1956 and Halwagi 1961,1962). The results of enclosure experiments by Kassas are very significant; within a year fenced land showed a tall grass cover of up to 80%, while the land open to goat grazing had only 5% cover; both soil erosion and a shift of desert were also observed.

In the river itself, colonisation takes place on sand banks bared by a falling level. A series of freshly exposed banks to old established islands was studied by Halwagi (1963), with a succession of plants. The same phenomenon exists on the White Nile, where the 'archipelago of islands'

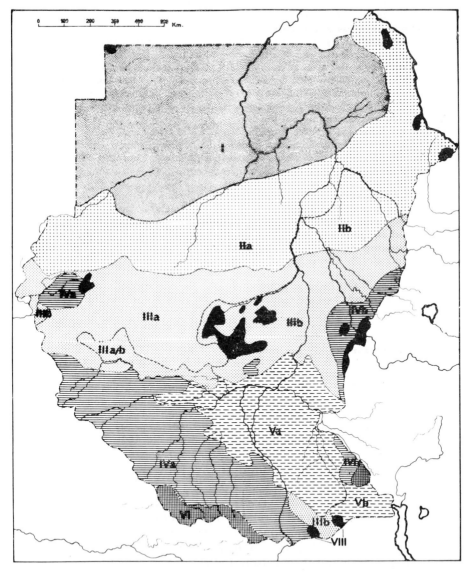

Fig. 11. Present vegetation zones in the Sudan; G. Wickens 1975. The area marked IIIa from the Northern end of the Nuba mountains southwest to the Sudan border should be included in area IVa.

83

Status after inundation

We must expect that the newly established riverain flora of lower Nubia will attain other features. The silt material which is necessary for the building up of the fertile embankments is no more available. Embankments if they ever exist would belong to a huge water reservoir rather than to a flowing river with active water currents.

Since the inundation of Nubia started in 1965, permanent human settlement is restricted to one locality only. A group of investigators started around the Abu Simbel temples a small experimental farm for the estimation of the water requirements of field crops under the new prevailing condition in Nubia. Cultivations in this farm are associated with some desert annuals and few common weeds which are introduced with the growing crops.

An early stage for the establishment of the riverain flora of Nubia was studied by the author during the last 2 years (1973–74) in the Daboud area which is located about 10 km south of the High Dam. That was the site of the old Daboud village and temple. The present Daboud island (3 km long, 200–400 m across) is a part of what was an elevated plateau east of the Daboud temple.

The few species recorded (about 15 species) belong to: the adjacent desert, the floating seeds and the aerially dispersed fruits of some inundated species.

Perennial vegetation of Daboud island is represented by a central core of large shrubs of *Salsola baryosama* (Schult.) Dandy and *Tamarix amplexicaulis* Ehrenb., which must have existed before the inundation; when this island was a part of the Libyan desert. A few specimens of the desert perennials: *Forskahlea tenacisima* L., and *Hyoscyamus muticus* L. were also recorded. At both sides of this perennial central core, there were zones of annual weeds which grew on the moist embankments of this island. Annual growths are arranged in longitudinal rows identical to the retreat of water level at regular intervals. The edges of Daboud island, where annual growths occur, remain submerged by the water during the summer months (high flood period). Late in autumn, seedlings of annual weeds appear on the moist ground where water retreated. This procedure was repeated resulting in a pattern of longitudinal rows of seedlings, the youngest of which was the nearest to the water. The recorded annual weeds were: *Glinus lotoides* L., *Senecio aegyptius* L., *Oligomeris linifolia* (Vahl) Macbr., *Rumex dentatus* L., *Amaranthus viridis* L., *Crypsis schoenoides* (L.) Lam.

Outstanding was the reed vegetation consisting of *Typha domingensis* Pers. and *Phragmites australis* (Cav.) Trin. ex Steud. developing in the shallow waters around the island and the opposite Khour El Berba. This is rather characteristic of the shores of a lake rather than of the banks of a river with active water currents. *Zannichellia palustris* L. and *Najas armata* Linn.f. were also growing among the weeds. The latter was not reported earlier from the southern waters of Egypt. It seems however to have been overlooked, and this species, and also another *N. minor* All. were recorded recently along the calm banks of the Nile in the area between Aswan and the High dam.

Acknowledgements

The author is greatly indebted for Prof. Dr. M. Kassas, Cairo University for reading and revising the manuscript.

REFERENCES

Abdallah, M. S., Sa'ad, F. M. & Abbas, A. 1972a. Taxonomical studies in the flora of Egypt. II. Natural flora of Egyptian Nubia before the construction of Aswan High Dam. Arab. Repub. Egypt, Agric. Research Centre Herb., Techn. Bull. 4: 1–83.

Abdallah, M. S. & Sa'ad, F. M. 1972b. Taxonomical studies in the flora of Egypt. III. Comparative studies in the natural flora of Egyptian Nubia. Egypt. J. Bot. 15: 265–281.

Ahti, T., Hämet-Ahti, L. & Pettersson, B. 1973. Flora of the inundated Wadi Halfa reach of the Nile, Sudanese Nubia with notes on adjacent areas. Ann. Bot. Fennici 10: 131–162.

Boulos, L. 1966. Flora of the Nile region in Egyptian Nubia. Feddes Repert. 73: 184–215.

Boulos, L. 1967. On the weed flora of Aswan, Egypt. Bot. Not. 120: 368–372.

El Hadidi, M. N. & Ghabbour, S. 1968. Floristic study of the Nile Valley at Aswan. Rev. Zool. Bot. Africaines 78: 394–407.

El Hadidi, M. N. & Kosinová, J. 1971. Studies on the weed flora of cultivated land in Egypt. I. Preliminary survey. Mitt. Bot. Staatssaml. München 10: 354–367.

Ghabbour, S. I. 1972. Flora of the Nile region at the Dongola Reach, Sudanese Nubia. Rev. Zool. Bot. Africaines 85: 1–29.

Pettet, A., Pettet, S. J., Cloudsley-Thompson, J. L. & Idris, B. E. M. 1964. Some aspects of the fauna and flora of the district around Wadi Halfa. Univ. Khartoum Nat. Hist. Mus. Bull. 2: 1–28.

6a. MAMMALS

by

K. WASSIF

Mammals of Egypt have been treated in a volume of the 'Zoology of Egypt' which appeared at the beginning of this century (Anderson, rev. by de Winton, 1902). Flower (1932) reexamined the mammalian fauna critically. Discounting 12 domesticated species, 3 'pests', and 4 marine species, there remain 76 species known at present. The largest group are the rodents, the bats and smaller carnivores. Human pressure has allowed mainly small species to survive, most of the larger ones have become extinct or are extremely rare. A leopard was shot at Mogharra in 1913, a cheetah on the Cairo-Alexandria desert road in 1967; the wild boar became extinct in the Delta about 1890; no records exist for the wild ass from the late 29th century onwards; rare are the gazelles in the desert, where motorised hunting parties operate; similarly with the ibex whose existence is doubtful. The hippopotamus, once almost a symbol of the Egyptian Nile valley, was by the 17th century so rare that the shooting of two near Damietta was recorded by the great Buffon in 1769 and described by Sonnini de Manoncourt (1799). The Arab chronicler Abdel Rahman ibn Hassan (known as El Jabarti) described the appearance of a hippo near Damietta in A.H. 1233 of the Muslim calendar, equivalent to 1815, if the story is true. By that time all memories of the animal had vanished in the population. Further upstream, Bruce saw one at Syene (Aswan) in 1786, but later only hearsay existed in Egypt. In Sudanese Nubia records exist for 1820 at Argo island, the latest record for Dongola is 1852; Baker found them frequently at the mouth of the Atbara river in the sixties of the last century. But gone are the times when Bruce reported an abundance at the confluence of the two Niles in 1771. At present the hippo is still present in the White Nile upstream from the Gebel Aulia dam and along the river up to the headwaters. In the Blue Nile it was never abundant, at present the two dams at Sennar and Roseires have closed the river; some survive in the Blue Nile Gorge, Lake Tana has none any more.

Remarks on some species should be added. The domestic buffalo was introduced in the Middle Ages probably from Mesopotamia. As draught animal and supplier of meat, milk and hides is has replaced largely other forms of cattle; these of ancient Egypt have been already described in a previous chapter; the cow was part of the cult as Hathor. The Egyptian mongoose *Herpestes ichneumon* and its useful role as destroyer of pests was recognised in ancient times with local cults e.g. at Heracleopolis in middle Egypt. The striped hyaena has become very rare since the time of

show clearly the wide expansive of the fauna in the southern part, the dependence of the presence upon the vegetation and very clearly the low nature of some species along the Nile valley. Here again there are differences between the eastern and western desert; the eastern part still harbouring a species extinct or not revealed in Egypt. Some species known from archaeological finds in the north are now confined to the Upper Nile swamp. (Damaliscus tiang and the Nile lechwe, and Kobus kob) seek the Sudanes...

This east African fauna and it is the dense part the wide extension of its fauna as the Nilo-headwaters in Eastern Africa. Some billion...

REFERENCES

Allen, G. M. 1939. A checklist of African mammals. Bull. Mus. Comp. Zool. Harvard 83: 1–763.

Sutherland J. 1982. Zoography of deep... Netherlands... et al. Mar. Biol. Rev. Embh...

Thomas, S. A. 1974. A census on the recent mammals of Egypt. Pop. Check. Sci. Found 10: 49–65.

Kock, D. 1969. Die Fledermaus-Fauna des Sudan. Abh. Senckenberg. Nat. Forsch. Ges. 521: 1–238.

Koopman, K. F. 1975. Bats of the Sudan. Bull. Amer. Mus. Nat. Hist. 154: 353–444.

MacKinnon, R. N. 1981. Distribution of wild mammals in the Sudan... Sudan style and fauna association. Sudan. Misc. Zool. Publ. Fauna no. 1. Khartoum.

Sexton, W. 1956. Mammals of the Sudan. Proc. U.S.A. National Mus. Senckenberg Inst. vol. 106, no. 3375. Washington.

Schouteden de Macheban, G. S. 1990. Travels in Upper and Lower Egypt. Transl. from French. Dawson. London.

Moreau, R. E. 1966. Several avifaunas on birds.

Hoogstraal H. (ed.), Egypt, 1961. II. Abh...

6b. BIRDS

by

J. RZÓSKA

This is a well explored group and modern sources of information exist; on these this account is based.

The Egyptian bird fauna comprises according to Meinertzhagen (1930) 73 species of which 45 live in the fertile Delta and half of these are palaearctic; he records 16 species for the desert and the rest for the oases and upper Egypt. Moreau (1966) lists 75 species, which he divides according to their environment: waterbirds with 27 species form 33% of the total, rapacious and scavengers are 13 (17%), groundbirds 5 (6%), the rest are small birds, Passerines and others. Of the total bird fauna 46% are palaearctic, 30% truly African.

The response of the fauna to the nature of the country is seen in the number of species associated with water. Moreau (in Meinertzhagen 1930) has also examined the birds of ancient Egypt on the basis of murals, hieroglyphic signs and mummified specimens and has recognised 90 species. The fauna was richer than at present and more African; four species of those found are now extinct in Egypt: the sacred ibis *Thresciornis aethiopicus*, *Phalacrocorax africanus*, *Rhynchops flavirostris* and *Struthio camelus* the ostrich. Of these the sacred ibis was seen until 1876, the cormorant vanished in 1875 from Lake Qarun in the Faiyum, the ostrich disappeared from the Egyptian desert more than 100 years ago. The shoebill *Balaeniceps* a true swamp animal seems to have disappeared during early historic times.

The Egyptian bird fauna is poor and of mixed character with affinities to the Maghreb, Sinai, Arabia besides the Palaearctis and Africa. This is due to the very impoverished vegetation, mainly confined to crops, palm groves and remnants of water vegetation. Faiyum is richer; the oases (Siwa, Kharga, Dakhla, Wadi Natrun) are according to Moreau poorer than Meinertzhagen thought and have only 4 resident species each at most.

Only 3 of the 25 desert species have not been recorded in Egypt, most are widespread across the Sahara; 2 species are known only from the eastern desert.

The contrast to the bird fauna southwards along the Nile is striking. The delta is the gateway for bird migrations which swell the numbers of birds seasonally; these penetrate as summer or winter visitors in enormous numbers.

Many observations on the bird fauna exist scattered in various publications e.g. The Sudan Notes and Records; Cave & Macdonald have collated these in a valuable book (1955). But the authors do not attempt to characterise the fauna zoogeographically and the following remarks are extracted mainly on the basis of distribution within the Sudan.

Cave & Macdonald list for the Sudan 76 families with 871 species and additional races, including non resident visitors: Moreau (1966) recognises only 497 species. Of these 22 families with about 150 species are associated with water habitats; the great and montonous Upper Nile swamps harbour 63 species; the north, arid and largely desert, has 34 species. The bulk of the fauna spreads across the country from the south to about 16° lat. N in conformity with the broad vegetational belts which run across similarly as the rain belts (see Fig. 10, 11). A number of distribution maps for some species shows clearly the narrowing of opportunities for bird life to the thin line of the Nile valley, along which elements of the Upper Egyptian fauna merges with that of the Sudan. Mass appearances of birds occur in the arid zone; sand grouse come in enormous flocks to the river in the morning as part of their daily routine. In the dry savanna further south weaver birds especially *Quelea* literally darken the sky and are a serious pest to crops and scrub forest. The great variety and richness of bird life from Khartoum upstream does not fail to impress even now. MacLeay (1959) travelling in September 1958 on the White Nile recorded 142 species including winter migrants from the Palaearctis and summer visitors from inner Africa; a colony of the shoebill was seen in the swamps. This is a large proportion of the whole fauna and shows the importance of the river.

The fauna is predominantly African (Ethiopian). Some remarks on the Ethiopian fauna are in the chapter on the Blue Nile Gorge (ch. 15).

Uganda and Kenya

The monumental work by Sir Fr. Jackson (1958) lists in 3 volumes the families and species of these two regions forming the headwaters of the White Nile. No general chapters are included but a wealth of information on almost each component of this rich African bird fauna.

It is Moreau (1966) who has assembled a revised general survey of the bird fauna of Africa. Of the 1481 species in the whole of the continent Egypt has 75, the Sudan 497 species, Uganda 495 and Kenya 565 species. Palaearctic migrants vary from 88 in the Sudan to 70 in Kenya; the phenomenon of migrations must be of relatively recent origin in view of the changes of climates both in Europe and Africa in the late Pleistocene.

REFERENCES

Cave, F. O. & MacDonald, J. D. 1955. Birds of the Sudan. Oliver & Boyd, Edinburgh-London.

Jackson, Sir Fr. 1938. Birds of Kenya Colony and the Protectorate of Uganda. vol. 1–3, Gurney & Jackson Publ.

MacLeay, K. N. G. 1959. Observations on birds of the Nile valley. Ann. Rep. Hydrob. Res. Unit Univ. Khartoum 6: 12–17.

Meinertzhagen, R. 1930. Nicoll's Birds of Egypt. Hugh Rees Publ., London.

Moreau, R. E. 1966. The Bird Faunas of Africa and its Islands. Acad. Press, New York-London.

REFERENCES

6c. AMPHIBIA AND REPTILES

by

FAWZI HUSSEIN

Egypt

The fauna of Amphibians and Reptiles of Egypt consists of 89 species; Marx (1968) distinguishes 93. They are distributed over the arid regions, the cultivated land and the aquatic habitat. Bufonid toads and a large array of geckonid, lacertid and agamid lizards, skinks and colubrid snakes are the most important herpeto-faunal elements reflecting the predominantly arid character of the country. The Egyptian fauna of these groups has a strong palaearctic character but 37% of the species occur also in the 'Ethiopian' African faunal region. Here the amphibian and reptilian species of Egypt are listed with short notes on their ecology.

AMPHIBIA

Anderson in his monumental 'Zoology of Egypt' vol. 1 (1898) named 5 species of which 3 may have been included from outside Egypt (2 species of *Phrynobatrachus* and one of *Hemisus*). Six species can at present be found: *Bufo regularis regularis* Reuss, *Bufo viridis viridis* Laurenti, *Bufo vittatus* Boulenger, *Bufo dodsoni* Boulenger, *Rana mascariensis mascariensis* Dumeril & Bibron, *Rana ridibunda* Pallas. Of these *Bufo r. regularis* is the most abundant and *Rana m. mascariensis* is found along the Nile valley deep into Africa. Marz (1968) has discussed affinities of this poor fauna.

Highly characteristic zoogeographically is the complete absence of tailed amphibians (newts, salamanders) as indeed from the whole of Africa, a clear sign of the influence of this continent. Savage (1973) has recently discussed the origin and the past and present distribution of the Amphibian Anura with a series of maps, which are revealing. For Africa except the mediterranean part and the north west some facts emerge: 1. Most Anura are concentrated south of the Sahara, which as one would expect is empty of amphibians; 2. Along the lower Nile valley only Bufonidae and Ranidae pierce the hostile Sahara and join their almost cosmopolitan wide expanse in the Palaearctic; 3. Tropical Africa is an ancient and rich centre of Anuran evolution; 4. The Phrynobatrachinae, Arthroleptinae, Hemisinae, Xenopinae, Pipinae, all sub-families of the Anura, reach the middle of the Sudan with some extensions to the Red Sea region.

Sudan

Only few papers deal with the amphibians of the Sudan and at present it is impossible to give a realistic number of species present; but with the remarks above one can safely assume a much richer fauna than in Egypt, especially as large parts of the country have not been explored.

Ethiopia (Abyssinia)

Here again the available literature is at present not sufficient to characterise it; a number of forms are endemic (Parker 1930) especially in the Highlands; the Blue Nile Gorge is a gateway for African forms as mentioned in the relevant chapter 15. But it is no doubt rich as compared with Egypt, with the isolation of the Ethiopian Highlands from the rest of Africa it may prove to be different.

Uganda and Kenya are rich in amphibian species. Loveridge (1957) lists 44 species for Uganda, Kenya with its greater variety of habitats has 65. Resuming, the amphibian fauna grows more abundant the deeper into Africa one goes; Egypt has only a weak link in this respect with the mother continent.

REPTILES

Egypt

Reptiles are well represented in the fauna, as one would expect from a group, which likes warmth and thrives in arid regions. There are 83 species or 88 according to Marx, consisting of 45 'lizards', 34 snakes, one chameleon, 2 Chelonia and the now extinct crocodile. Families and numbers of their species are: Geckonidae 14, Agamidae 9, Lacertidae 12, Scincidae 8, Varanidae 2; of the snakes: Typhlopidae 1, Leptotyphlopidae 2, Boidae 2, Colubridae 20, Elapidae 3, Viperidae 6. As to numbers of individuals encountered, 8 lizards are most common besides two toads mentioned above. Compared with the older lists by Anderson and by Boulenger recent investigations (Hussein *et al.*) have increased our knowledge of the fauna and their ecology. As member of an USA medical team Marx (1968) has collected intensively around Cairo and few other locations; his $3\frac{1}{2}$ thousand specimens represent about 78% of the known fauna. Flower (1933) roamed about more widely and collected at Wadi Natrun, the Faiyum, Alexandria, the Delta, Giza and Cairo. He found 24 species in Upper Egypt from Giza to Aswan and 14 from Aswan to Wadi Halfa. Of the whole fauna only 14 species are water bound in a varying degree (3 skinks, 8 snakes, the Nile monitor now extinct in lower Egypt, the turtle Trionyx and the crocodile similarly absent now; true desert forms especially the Agamidae and some snakes number 20 species. Between these two habitats live the other forms. There are 4 poison snakes (*Naja haje haje, Naja nigricollis nigricollis, Cerastes cerastes* and *Cerastes vipera*).

More than 60% of the fauna have palaearctic affinities with the Mediterranean North Africa, the Levantine, and even southern Russia. The rest points to Africa.

Sudan

Only the snakes are reasonably well known; very early guides to snake recognition have been issued by government agencies for medical reasons. Corkill (1935) listed 39 species in 6 families from Wadi Halfa down to the southern border of the Sudan. Later collections (Loveridge 1955) increased this number to 44 species. Most of the species are African.

Uganda and Kenya

According to Loveridge (1957) the herpeto-fauna of these two headwater regions of the Nile consists of:

	Uganda	Kenya
Lizards and chameleons	41	92
Snakes	69	89
Caecilia	—	3

Water turtles and tortoises are present and the Nile crocodile. Recently Pitman, who was game warden in East Africa for many years, has written a 'Guide to the snakes of Uganda' (1974).

Special consideration must be devoted to the Nile crocodile; its disappearance from Egypt is well documented, its progressive extinction along the Nile valley is rapid.

THE NILE CROCODILE

It ranked high in ancient Egyptian records, it was worshipped in some parts of Egypt. The story of its disappearance can be pieced together from many observations made by travellers who began to visit Egypt and the Nile valley in increasing numbers from the end of the 18th century onwards. Dates and sites could be compiled extensively from 1786 to the end of the 19th century, from Beni Suef, about 140 km south of Cairo to Abu Simbel in Egyptian Nubia. Besides the increasing intensity of agriculture and population, the erection of barrages and dams reduced opportunities for survival. Only stray specimens were left; Flower (1933 and some earlier notes) speaks of one seen at Rahmaniya on the Rosetta branch, 120 km north of Cairo. He also records some crocodiles at Korosko as late as 1920.

Further south, in Sudanese Nubia, travellers reported more; even in 1898 troops of Kitchener's expeditionary force were warned against crocodiles in the Ambukol – Korti region. In the early 20th century Flower (1933) records 'fairly numerous' crocodiles at Wadi Halfa, Dongola and Karima (IV cataract); he hears complaints of a local sheikh about human casualties in 1920. More recently only sporadic encounters are known such a collision of a canoe with a swimming crocodile near Atbara in 1933 (R. Hill pers. communication). Around 1950 a small colony lived in the Sabaloka gorge north of Khartoum, few middle sized were regularly seen at Gerif a small village on the Blue Nile near Khartoum. Gone were the days when Bruce saw an 'abundance' at Halfaya near the confluence of the two Niles about 1771. A very exact record on the Blue Nile has been left by Flower (l.c.) who travelled from Khartoum to Roseires five times in 1905–1910. Up to Wad Medani he saw averagely 7 crocodiles per day, in the Sennar region 27–42 in each day of travel, at Singa 123 inone day, in the Roseires region 37–64 animals. Two years later he still records 41 crocodiles between Singa and Sennar. This was before the Sennar dam closed the river. Hammerton, who worked here recently, records only few specimens in the Fazughli Gorge. Morris (ch. 15) reports on crocodiles in the Blue Nile gorge and on the hunting expeditions organised from the Sudan. There are no crocodiles in Lake Tana; it is not certain why. Of the three temporary rivers entering the Nile system from the east, the Atbara was once rich in big fauna and crocodiles are probably still encountered in its lower course and certainly in its Ethiopian tributaries. The Rahad and Dinder are poor; Grabham found a nest in the Rahad in 1909 and reported this to 'Nature'!

For the White Nile south of Khartoum every traveller in the 19th century wrote about the number of crocodiles increasing towards the south. In the course of time these numbers dwindled; but on the whole this was a favourable environment with no dams and large inundations over the flat shores. The Gebel Aulia Dam closed the river in 1936 and this changed the environment profoundly.

The great Upper Nile swamps once abounded in animal life; Flower (1933) speaks of the large crocodiles he saw during his travels in the beginning of the century. Professional hunters even from outside the Sudan decimated the stock. In 1951 skins are recorded as important export item to the number of 26,000 and this went on. Only patchy distribution could be seen in 1953–1956 in some places.

The most thorough scientific investigations on the status of the Nile crocodile was made by Cott (1961, 1969) in Uganda. In the last 30 years an enormous destruction by professional hunters brought this species to near extinction especially in Lake Victoria. Elaborately equipped hunting expeditions were organised; one of these shot 1,000 crocodiles in six weeks of 1954. In 1957 over 60,000 skins were sold from the whole of East Africa. No animal species can withstand this toll, especially one with a very delicate breeding habits.

Until recently a refuge existed in the national park at the Murchison Falls. Cott studied this area for several years and left us a careful account of the ecology and physiology of the crocodile; from this monographic study some items are here given.

Crocodiles have a diurnal rhythm of activities; they spend the night mainly in water, the day on land and this is necessary for their thermo-regulation. Growth and age records show that big crocodiles of 21 feet (about 7 m) are 100 years old. A very clear change of diet occurs during their life; young animals from 0.3 to 1.5 m long live on invertebrates, insects and molluscs mainly, later they switch to fishes and less commonly to birds, reptiles and mammals. Only in 2 out of 500 specimens investigated in Uganda were there any human remains. Ennemies like baboons, hyaenas, birds and above all the Nile monitor *(Varanus niloticus)* take a heavy toll of eggs and young. Parasites include the Tsetse fly which infects the blood with *Trypanosoma grayi* and the haemogregarine *Hepatozoon pettiti*; besides, 3 species of leeches and two of Nematodes have been found. Widely commented upon since antiquity are birds associated by some form of commensalism with crocodiles basking on the shores. Cott singles out two species flitting around and picking out food from the gaping mouths of crocodiles: *Hoplopterus spinosus*, probably the 'trochilos' of Pliny, and the sandpiper *Actitis hypoleucos* whose warning cries of danger are readily responded to by their hosts.

The breeding season is in December to January and from August to October; the young hatch in between the two main but ill defined rainy seasons in November and April/May. The number of eggs of one nest varies between 51–62, of which a large number is destroyed by predators. Disturbance by humans is also fatal as the mothers flee and the young cannot get out of the sand pits by themselves. Every year a number of breeding grounds are found abandoned. Cott estimated the number of eggs produced along the 28 km of the Murchison/Victoria Nile at 10,000 per year. Of this number a large proportion could be saved for conservation by artificial breeding. There is a large scope for crocodile farming and experiences already exist in South Africa and in the Tchad area. This could satisfy commercial demands and save the species. Cott concludes his study by stating firmly that the Nile crocodile is a useful member of the water community; his detailed diagram of food relations bears this out convincingly.

In general, here we have another picture of withdrawal and decimation due to lack of understanding the resources of nature. In various places along the Nile valley the problem of intelligent management by conservation and farming could help to preserve a valuable species. There are surprising indications that in the great High Dam basin of Aswan stretching deep into the Sudan some footprints of crocodiles have been observed; the great artificial lake is now banked on both sides by the desert, the population has been removed and only fishing camps exist.

A digression on the spread of the Nile crocodile outside the Nile valley may be included. Lake Chad has crocodiles as part of its generally 'nilotic' faunal character, described by Greenwood (ch. 7). Attempts are being made to farm, as mentioned by Le François (Rapp. Annuel, ORSTOM 1973).

Much further into the Sahara relict survival of the Nile crocodile was reported to exist until 50 years ago in the Iherir area of the Tassili. Some of the water bodies remaining there of the former aquatic network contain isolated nilotic fishes, and contained the Nile crocodile in rudimentary size; the last was apparently shot in 1924 according to Lhote (1958 p. 201–2.). Staudinger has reported interesting details on 'Krokodile in der Inner-Sahara und Mauritania' (Sitz. ber. Naturf. Ges. Berlin no. 4–7, 1929).

REFERENCES

Anderson, J. 1898. Zoology of Egypt. vol. 1. Reptilia and Batrachia. London.

Corkill, N. L. 1935. Notes on Sudan Snakes. Sudan Govt. Nat. Hist. Mus. Khartoum, Publ. no. 3.

Cott, H. B. 1961. Scientific Results of an enquiry into the ecology and economic status of the Nile Crocodile, in Uganda and Northern Rhodesia. Transact. Zool. Soc. London 29: 211–356.

Cott, H. B. 1969. Further observations on the status and biology of the Nile crocodile below Murchison Falls. U.K., I.B.P. report.

Flower, S. S. 1933. Notes on the recent Reptiles and Amphibians of Egypt. Proc. Zool. Soc. London 1933: 735–851.

Hussein, M. F. 1955. Ecological studies in Anquabiya region with special reference to thermal relationships of some Egyptian desert reptiles. Ph.D. Thesis, Cairo Univ.

Hussein, M. F. 1961. Some aspects of the ecology of Zamenis diadema with special reference to thermal relationships. 4th Arab Sci. Congr., Cairo, p. 66.

Hussein, M. F., Boulos, R. & Badry, K. S. 1969. Activity of the lizards Chalcides ocellatus and Agama stellio with special reference to light and temperature. Proc. Zool. Soc. U.A.R. 3: 67.

Hussein, M. F., Boulos, R. & Badry, K. S. 1974. Activity of some Egyptian reptiles with reference to light, temperature and humidity. Bull. Fac. Sci., Cairo Univ., 46 (in press).

Khalil, F. & Hussein, M. F. 1961. Ecological studies in the Egyptian deserts. II. Notes on some lizards inhabiting Angabiya region. 4th Arab Sci. Congr., Cairo, p. 25.

Khalil, F. & Hussein, M. F. 1963. Ecological studies in the Egyptian deserts. III. Daily and annual cycles of activity of Uromastix aegyptia, Agama pallida and Chalcides sepoides with special reference to temperature and relative humidity. Proc. Zool. Soc. U.A.R., 1: 93.

Lhote, H. 1958. À la découverte des fresques, du Tassili Arthaud.

Loveridge, A. 1955. On snakes collected in the Anglo-Egyptian Sudan by J. S. Owen Esq. Sudan Notes and Records 36: 1–20.

Loveridge, A. 1957. Check list of the Reptiles and Amphibians of East Africa, (Uganda, Kenya, Tanganyika, Zanzibar). Bull. Mus. Comp. Zool. Harvard 117: 153–362.

Marx, H. 1968. Checklist of the Reptiles and Amphibia of Egypt. Spec. Publ. U.S.A. Naval Med. Res. Unit no. 3, Cairo.

Parker, H. W. 1930. Report on Amphibia collected by Mr. Omer Cooper in Ethiopia. Proc. Zool. Soc. London 1930.

Pitman, C. R. S. 1974. A guide to the snakes of Uganda. Rev. ed. Codicote, Wheldon & Wesley, Glasgow.

Savage, J. M. 1973. The geographical distribution of frogs, patterns and predictions. In: J. L. Vial (ed.) Evolutionary Biology of the Anurans. Contempor. Res. on Major Problems. Univ. of Missouri Press.

6d. INVERTEBRATES

by

J. RZÓSKA

In general the invertebrate fauna of Nile areas is not sufficiently known to allow for zoogeographical reflections or conclusions. In the wake of explorers of Africa came also botanists and zoologists. The flora and the vertebrates attracted more attention than invertebrate groups. Yet a number of 'fact finding' expeditions, e.g. a Swedish and an Austrian led by Werner, supplied a number of papers on terrestrial and some aquatic groups; collections in museums allowed later for early monographs. But these are out-dated and to examine them critically would be a huge undertaking. What has been done in the field of terrestrial fauna of the Sudan and Egypt reflects the dry character of most of the Nile area e.g. in a number of insect groups.

Desert Zoology

The contrast between the Nile Valley s. str. and the arid lands gripping the river habitat on both sides for almost three thousand km is an important characteristic and must be referred to. Botanists and zoologists who have written about the central and northern Sudan and Egypt have stressed the 'arid zone' character of flora (ch. 5) and fauna. Recently a number of papers and books have appeared by Cloudsley Thompson and collaborators, e.g. 1962, 1963, 1965 and Pettet *et al.* (1964), who worked at the University of Khartoum. These papers dealt with the composition and adaptation of faunal elements to the conditions in the eastern Sudan, Khartoum and Wadi Halfa. There is a severe selection of groups and species fit to live under desert conditions with physiological adaptations and above all a concentration in microclimatic habitats providing shelter.

Yet, a recent book on 'Desert Biology' (G. W. Brown 1968), includes a chapter on 'Desert Limnology'. This contains a.o. a summary of a study on rainpools carried out at Khartoum (Rzóska 1961). These pools appear only after the strongly seasonal summer rains; under the climatic conditions prevailing, an average of 160 mm rain and high temperatures, these pools last only from a few days to 2 or 3 weeks. They contain a fauna extremely rich in individuals from a very selected group of animals: *Triops*, Anostraca, Conchostraca, two copepods *Metacyclops minutus* (Claus) and *Metadiaptomus mauretanicus* Kiefer & Roy, *Moina dubia* de Guerne & Richard, and two rotifers of 'Asplanchna' and 'Pedalion' type. Their rapid development from drought resisting eggs or stages to full maturity in a few days and their survival in high water and even higher soil

111

temperatures (up to 80 °C) are remarkable. Cloudsley-Thompson has reported on the lethal temperatures of *Triops* and on orientation responses of two species of Euphyllopods. Carlisle (1968) has experimentally shown the survival of *Triops* eggs under temperatures a little below boiling point and the necessity of high temperatures to break the egg diapause.

For prehistoric water bodies in the desert see chapter on Palaeoecology.

The alluvial soil

Between the desert and the river stretches a strip of alluvial soil of varying width, which is in some parts occupied by cultivation, irrigated or wetted by the Blue Nile flood. This strip is much richer in plants and animals than the desert, as described by Pettet *et al.* (1964). Two papers deal more specifically with the transient habitat of wet silt. Brook (1952 and 1956) discovered at Khartoum the remarkable terrestrial Chaetophoraceous alga *Fritschiella tuberosa* Iyengar, which was described first from wet soils in various parts of India; the Khartoum find seems to be the second continent for this alga. It appears here from August to October in the gradually drying silts after the fall of the Blue Nile flood. Another observation describes the succession of terrestrial invertebrates living in this habitat (J. G. E. Lewis 1961/63). This is a fauna rich in individuals composed of Carabid and Tenebrionid beetles and small Orthoptera near to the wet shore, followed inland by Lycosid spiders, Solifugae and scorpions. These groups appear and disappear with the flood rising or falling according to their peculiar requisites. The stark transition of microclimatic conditions from wet to inland conditions is worth quoting: at noon the relative humidity over wet silt never fell below 100%, 10 m inland it was 10%. Temperatures showed a similar contrast of 39.5 °C to 62.5 °C on the surface of the silt.

Aquatic Invertebrates

For Protozoa, Sponges, Hydrozoa, Bryozoa, Nematodes, Hirudinea and many insect groups the available references are scattered in a number of papers too specialised to provide for general conclusions without a detailed study. For these groups no attempt will therefore be made. Better investigated are the Mollusca, the Rotifera, Crustacea and the aquatic Hemiptera. The Oligochaeta treated in this book by Ghabbour (ch. 25) are not well documented for some stretches of the Nile system. Some of the groups owe their better exploration to particular circumstances such as the presence of resident biologists (not 'expeditions') or to the use as medical or archaeological indicators or simply to the zeal of ardent collectors.

Mollusca have been intensively investigated in Lake Victoria and other Uganda waters by Mandahl-Barth (see ch. 25), those of L. Tana

112

by Bacci (ch. 14); the nilotic mollusc fauna was mainly investigated for the presence of potential vectors of schistomiasis (ch. 28) and indicators of palaeoecological conditions (ch. 4a and fig. 12) It seems that climatic changes account only for some minor changes in the aquatic nilotic fauna in the last 20,000 years. Rotifera and some Crustacean groups form an important part of the zooplankton (ch. 24). The very varied shore fauna was investigated best in Uganda swamps and in the Upper Nile swamps (ch. 11 and 12); crabs belong mainly to the benthos (ch. 25). Parasitic Crustacea, mainly from fishes, are regarded by Fryer in a number of papers as zoogeographically significant.

The aquatic Hemiptera have found recently an overall treatment by R. Linnavuori (1971); his collections extend to many parts of the Sudan, Ethiopia, Somalia and 3 sites in South Yemen. Included in this paper are the species collected by Rzóska in the Upper Nile swamps and determined by Poisson (1951). Previously Hutchinson has attempted to review the zoogeography of this group in the African continent. Linnavuori has added 52 new taxa to the previously known fauna, mainly in the Sudan, which was hitherto a 'white spot'. Altogether he distinguishes 152 species and subspecies in the area investigated, many species are widespread in Africa. Hutchinson discusses the mixing of palaearctic and African ('Ethiopian') elements of the Hemiptera. This is apparent especially in the Hoggar mountains; North Africa (Tunis, Algeria and Morocco) is predominantly palaearctic, but Egypt has a greater proportion of African species, though records of the Hemipteran fauna in Egypt are

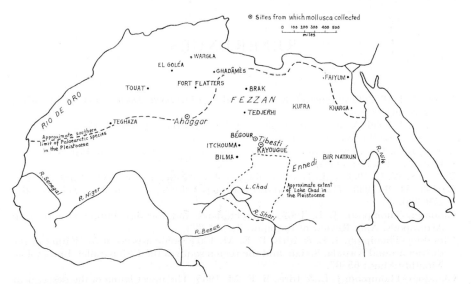

Fig. 12. Freshwater molluscs in the Sahara, spread of palaearctic species in the Pleistocene; Note: extent of Lake Chad in the Pleistocene; from Sparks & Grove 1961, J. Linn. Soc. London 44.

113

old and our knowledge is limited. The Nile is the 'only channel' to the spread of African species into North Africa. Abyssinia has two distinct faunas in the lowlands and in the highlands; the fauna is partly endemic with African intrusions (Brown 1965). Uganda shows endemic species in the investigated Kampala region; the great collections by Hancock and Hopkins are probably the basis for Hutchinson's views.

Finally some peculiar finds should be mentioned: *Cordylophora caspia* Pallas (= *lacustris* Allman) has been found in the Upper Nile swamps (Rzóska 1949) with the nearest site where it flourishes in the Birket el Qrun of the Faiyum and lake Edku of Egypt. The endoproctan *Urnatella* has been also found in rich encrustations on plant debris in the swamp region of the Sudan in 1954. A zoogeographical puzzle has been created by an 'Argulus' like specimen found by Rzóska in a sample of Nile plankton. Fryer and other specialists examined it and after considerable doubts it was declared to be a species of *Lichomolgus*, entire-marine and living in a variety of invertebrates including Tunicates and Coelenterates. (Fryer in litt.).

Of great interest are the two studies by I Thornton on 'communities' living in the floating plant *Pistia stratiotes* and that in papyrus heads forming a succession of terrestrial groups of considerable intricacy (see ch. 12).

REFERENCES

Brook, A. J. 1952. Occurence of the terrestrial alga *Fritschiella tuberosa* Iyengar in Africa. Nature 164: 754.

Brook, A. J. 1956. A note on the ecology of the terrestrial alga *Fritschiella tuberosa* in the Sudan. New Phytologist 55: 130–132.

Brown, D. S. 1965. Freshwater gastropod mollusca from Ethiopia. Bull. Brit. Mus. Nat. Hist. D. 12 no. 2: 39–94.

Brown, S. W. Jr. 1968. (ed.) Desert Biology. vol. 1. Academic Press, New York-London.

Carlisle, D. B. 1968. *Triops* (Entomostraca) eggs killed only by boiling. Science 161: 279–280.

Cloudsley-Thompson, J. L. 1962. Microclimates and the distribution of terrestrial Arthropods. Ann. Review of Entomol. 7: 199–222.

Cloudsley-Thompson, J. L. & Idris, B. E. M. 1963. Some aspects of the Fauna of the district around Kassala, Sudan, and the region south of the 13th parallel. Entomolog. Monthly Mag.: 65–67.

Cloudsley-Thompson, J. L. & Idris, B. E. M. 1965. The insect fauna of the desert near Khartoum; seasonal fluctuations and the effect of grazing. Proc. R. Ent. Soc. Lond. A 39: 41–46.

Cloudsley-Thompson, J. L. 1967. Animal Twilight. Foulis USA Dufour.

Cloudsley-Thompson, J. L. 1965. The lethal temperatures of *Triops granarius* (Lucas) Branchiopoda: Notostraca. Hydrobiologia 25: 424–425.

Cloudsley-Thompson, J. L. 1966. Orientation responses of *Triops granarius* and *Streptocephalus* ssp. Hydrobiologia 27: 33–38.

Hutchinson, G. E. 1933. Zoogeography of the African aquatic Hemiptera in relation to past climatic changes. Int. Rev. Hydrob. & Hydrogr. 28: 435–468.

Lewis, J. G. E. 1961–3. The ecology of the riverain invertebrate fauna of the Blue Nile at Khartoum. Hydrob. Res. Unit Ann. Rep. 9, 10: 17–18.

Linnavuori, R. 1971. Hemiptera of the Sudan, with remarks on some species of the adjacent countries. 1. The aquatic and subaquatic families. Ann. Zool. Fennici 8: 340–366.

Pettet, A. et al. 1964. Some aspects of the fauna and flora of the district around Wadi Halfa. Univ. of Khartoum Nat. Hist. Mus. Bull. no. 2: 1–28.

Poisson, R. 1951. Mission A. Villiers. Hemiptères Cryptocérates. Bull. Inst. Français Afrique Noire 13: 131–139.

Rzóska, J. 1949. *Cordylophora* from the Upper White Nile. Ann. Mag. Nat. Hist. Lond. ser. 12. 2: 588–560.

Rzóska, J. 1961. Observations on tropical rainpools and general remarks on temporary waters. Hydrobiologia 17: 265–286.

Swedish Expedition to Egypt and the White Nile under the dir. of L. A. Jägerskiöld, Uppsala 1904–1929; numerous contributors.

Werner, F. 1905. Ergebnisse d. zoolog. Forschungsreise Dr. F. Werner's in den ägyptischen Sudan und nach Nord-Uganda. Sitz. Ber. Kais. Akad. Wiss. Wien from 1905 onwards; numerous contributions.

and citing Furon (1967) and Chumakov (in Butzer & Hansen 1968) proposes that Nubia had been a drainage axis since Precambrian and that the Nubia Nile acquired its present course when it reached crystalline rock in the Oligo-Miocene and cut it deeply in Upper Miocene, which was a period of general uplift in Egypt and marine regression. There is evidence that the Red Sea mountain chain was present at least as early as Middle Miocene. Maley (1970) further suggests that the Nubia Nile already drained the central Sudan in the Pliocene (based on Chumakov in Butzer & Hansen 1968) and that the presence of *Podocarpus gracilior* pollen (associated with Ericaceae, Acanthaceae and Pedaliaceae pollen) from Lower Quaternary (Calabrian) to Post-Tyrrhenian submarine deltaic deposits is proof of a connection between the East African highlands and the Mediterranean all through the Quaternary. He relates this with a probable Tertiary – Quaternary phase of tectonic activity in Nubia which was an extension of the Rift Valley tectonics and which rejuvenated the whole relief of Nubia. Maley (1969) suggests that this same phase brought the masses of Lake Victoria waters to the Sudan depression where they found their way to the Mediterranean without much delay.

This theory agrees with the conclusions of Ball (1939) concerning the presence of a drainage axis in Nubia starting in the Eocene, but contradicts the more widespread theory of two separate Niles in Egypt and the Sudan which became connected some time between 32,000 to 50,000 years ago, although Butzer & Hansen (1968) do not exclude the possibility of earlier tributaries of the Egyptian Nile originating in the Sudan and presume that any previous sediments were removed by the massive subsequent floods of the modern Blue Nile. De Heinzelin (1966) gives evidence of a diluvial Nile flood 21,000 years B.P. at an aggradation absolute height of 157 m, never reached later but lower than earlier aggradation phases antedating the Upper Pleistocene. A number of major wadis in Nubia must have been active in the Tertiary-Quaternary, such as Allaqi, Amur, Mogaddam and Melek, and probably the Atbara. These wadis are part and parcel of the extensive wadi system of the Red Sea mountain chain. They must have maintained the Nubia Nile flowing during the whole Pleistocene and even during the dry Pliocene (when it flowed into the marine gulf at the locality of Aswan). Butzer & Hansen (1965, 1966) identify five or six periods of increased stream competence in southern Egypt covering the time range of the Early and Middle Pleistocene. In addition to these intervals of accelerated wadi and geomorphic activity, there was repeated formation of deep, red palaeosols. There can be no doubt that periods of accelerated wadi activity in southern Egypt were of even greater activity in the larger wadis of northern Sudan.

But the main controversy raised by Maley's theory is that, because of pollen evidence alone, he ascribes the outflow of the Lake Victoria waters to Early Pleistocene, while it had been ascribed by de Heinzelin (1964) to Middle Pleistocene. The *Podocarpus* pollen and its associates

118

Table 1. Oligochaeta of zoogeographical importance in the Nile basin and adjacent regions (from various sources, mainly Brinkhurst 1966 and Brinkhurst & Jamieson 1971).

Species	Egypt		Sudan		Ethiopia	Uganda	Kenya	Tanzania
	Siwa Oasis	Nile valley and Delta	White Nile system	Blue Nile				
Aeolosomatidae								
Aeolosoma headleyi		+	+(1)					
Ae. hemprichi		?						
Naididae								
Allonais paraguayensis	+		+				+	
All. pectinata			+			+		
Branchiodrilus hortensis								
Chaetogaster crystallinus			+			+		
Dero cooperi			+			+		
Dero digitata			+					
Dero (Aulophorus) pigueti								
D. (A.) flabelliger								
Paraneis litoralis		+(2)						
Pristina jenkinae						+	+	
P. longiseta sinensis							+	
P. proboscidea								+
Tubificidae								
Aulodrilus pigueti			+					
Opistocystidae								
Opistocyste sp.			+					
Alluroididae								
Alluroides brinkhursti abyssinicus					+			
A.b. brinkhursti						+		
A. pordagei							+	
A. ruwenzorensis						+		
Syngenodrilus lamuensis							+	
Eudrilidae								
Chuniodrilus ghabbouri		+						
Eminoscolex barnimi				+				
Eudrilus eugeniae					+			
Acanthodrilidae								
Acanthodrilus ragazzi					+			
Megascolecidae								
Gordiodrilus dominicensis			+					
G. habessinus					+			
G. niloticus			+					
G. pampaninii (3)								
G. paski								+
G. siwaensis	+							
G. zanzibaricus (4)								+
Nannodrilus staudei		+						
Pygmaeodrilus affinis		+						
Pygmaeodrilus sp.	+							

119

Table 1 *(continued)*

Species	Egypt		Sudan		Ethiopia	Uganda	Kenya	Tanzania
	Siwa Oasis	Nile valley and Delta	White Nile system	Blue Nile				
Glossoscolecidae								
Alma emini			+			+	+	+
A. nilotica		+	+(1)					
A. stuhlmanni		+				+		
Callidrilus ugandaensis						+	+	
Glyphidrilus stuhlmanni								+
G. sp.					+			
Lumbricidae								
Eisenia rosea var.?	+							
E. r. f. bimastoides		+						
Allolobophora jassyensis var. *orientalis*		+						

1. North of Khartoum.
2. Fayum (Lake Qurun).
3. Tarhuna, Tripolitania, Libya.
4. Also in Fada, Ennedi, northern Tchad (not in Selima Oasis).

could have been incorporated in the submarine deltaic sediments from the headwaters of the Sudanese tributaries of the Nubia Nile in the Red Sea mountains, the northern and western fringe of the Ethiopian highlands or from the highlands of southern and southwestern Sudan. *Podocarpus*, though characteristic of East African highlands, exists in the mountains of southern Sudan and could extend its range with any cooling and/or any increase of rainfall resulting in a slight lowering of the tree-line.

Zoogeography usually has to base its conclusions on sound geological evidence and may have to wait until this evidence is produced, but it is also quite possible for zoogeographic evidence to benefit geology by favouring one theory over another. The distribution of oligochaetes in the Nile Basin presents some zoogeographic puzzles which can be solved only by postulating a long history of Pleistocene Sudano-Egyptian Nile connection, thus corroborating the ancientness of the Nubia Nile.

The present distribution of aquatic oligochaetes in Egypt (Table 1) shows that there have been at least two waves of migration from African sources. The first wave brought *Allonais*, *Pygmaeodrilus* and *Gordiodrilus* to Siwa Oasis and further to Tripolitania *(G. pampaninii)*. *Allonais paraguay-ensis* of Siwa is related to conspecific varieties of the Sudd swamps, Kenya and South Africa. *Pygmaeodrilus* sp. is outside the distribution area of the genus from the Lake Albert region to southern Africa. Although *P. affinis* was found near Cairo (Brinkhurst & Jamieson 1971), this record is more likely to belong to a later wave of migration than the one which brought the genus to Siwa. The nearest locality for *P. affinis* is in Uganda (Gavrilov 1967). *Gordiodrilus siwaensis* is related to other members of the genus in Tanzania, West Africa and Ethiopia but the genus is represented in southern Sudan which is believed to be where the original *Nannodrilus-Gordiodrilus* stock sprang (Khalaf & Ghabbour 1968a).

Jamieson (1957) believes that the stem-form of the genus *Pygmaeodrilus* segregated into two groups when the Albertine rift was formed and the Victoria Nile outflowed via the north end of Lake Albert. According to earlier theories (de Heinzelin 1964) this happened probably in the Middle Pleistocene, and according to Maley (1970) in the Early Pleistocene. Since the Siwa depression was formed in Late Pliocene and Early Pleistocene by wind erosion and other factors during an extremely arid phase (Pfannenstiel 1953), the presence of aquatic oligochaetes in the Oasis cannot be from a pre-existing Pliocene fauna (the land was extremely dry and the Nile Valley was a marine gulf). It is possible only by migration of species from further subsequently wet habitats during a pluvial period when rainfall could render the soil of the Mediterranean coast sufficiently wet for the crossing. The northernmost Sahara was wet enough for human habitation in Early and Middle Palaeolithic as evidenced by the presence of implements as far as 200 miles south of the Mediterranean coast for the Early Palaeolithic (McBurney 1960) and the edge of the Sand Sea for the Middle Palaeolithic (Murray 1951). The degree of humidity of the early Würm moist phase, while comparable with that of the Early to Middle Pleistocene pluvials, was not matched again in Late Würm and Holocene times (Butzer & Hansen 1965 and 1966). It is unlikely therefore that migration of aquatic oligochaetes to Siwa took place either prior to the Pleistocene or in the post-Pleistocene, but in the last major Pleistocene pluvial phase over the Mediterranean coast. This major phase should be one after the northward outflow of the Victoria Nile (whether in Early or Middle Pleistocene) and it should also be the last one, because if it was not the last one, we should find members of the second wave of migration, now found in the Delta, in Siwa too. This second wave must have reached the Delta at a subsequent time. The absence of the Siwan group from the Delta, although the Nile was their original route of migration, may be due to a temporary submergence of the Delta during a part of the early or mid-Holocene (Ball 1939). When the Delta re-emerged, it was isolated from Siwa by desiccation of the coast (Khalaf & Ghabbour 1968b).

Members of the second wave now found in the Delta consist of *Chuniodrilus ghabbouri*, *Alma stuhlmanni*, *Nannodrilus staudei* and *Pygmaeodrilus affinis*. The genus *Chuniodrilus* was known from West Africa only, until Jamieson (1969) described the endemic species *Ch. ghabbouri* from Dahshur, near Cairo. The genus *Alma* is represented in Egypt by two seemingly unrelated species, *A. nilotica* and *A. stuhlmanni* (Jamieson & Ghabbour 1969). The first is a typical dweller of the mud and silt flats of the Nile banks from the Delta to Khartoum (and also in Bahariya Oasis, new record) while the second is the most widespread of the 13 *Alma* species inhabiting the Gambia, the Niger, the Congo, Lake Victoria and the Nile system. *A. stuhlmanni* occurs in West Africa, Uganda, Zaire and Tanzania (Brinkhurst & Jamieson 1971). Recent evidence indicates that the ancestor of *Alma* is provenant from a Tertiary Palaearctic source and that *A. nilotica*, with gills attesting to a long aquatic history, is very near this ancestor (Brinkhurst & Jamieson 1971, pp. 156–159). That *A. nilotica* occurs in Bahariya Oasis and has a range extending to Khartoum only shows that it has a different history from *A. stuhlmanni* and other members of the second wave. *A. stuhlmanni* has a restricted occurrence in irrigation canals west of Cairo. Jamieson (in Brinkhurst & Jamieson 1971, p. 152), who considers that collection of oligochaetes in Africa has been sufficiently intensive and widespread to suggest that patterns of distribution of *Alma* are real, though much detail remains to be filled in, refutes the view that limnic species are of no value in zoogeographical studies by evidence of the remarkable degree of local endemism shown by *Alma*. *A. emini* and *Pygmaeodrilus affinis* cohabit the drainage basins of Lakes Albert and Victoria but the first species stops short at Tonga while the second occurs near Cairo, confirming that the Nile does not provide a free passage for whichever aquatic oligochaete. An ancestor of *A. nilotica* could travel upstream and give rise in the reaches of central Sudan to a divergent number of species which spread in the rivers of central Africa and from there to West and East Africa. The emergence of *A. emini* as a distinct species is thought by Jamieson (in Brinkhurst & Jamieson 1971, p. 789) to have occurred from an ancestral population which was divided by the Albert-Tanganyika rift, that on the Congo side giving rise to *A. pooliana*, and that on the east to *A. emini*, which is now found in the great lakes and southern Sudan. The widespread general distribution of *A. stuhlmanni* in contrast with the limited distribution of *A. emini* may be explained by the eurythermal character of the former. Among the four members of the faunule consisting the second wave of migration, one, *Nannodrilus staudei* is endemic in the Delta and Kharga Oasis, thus showing a less restricted distribution than the other three. The genus is represented in West Africa (*Nannodrilus africanus* Beddard 1894).

It is thus evident that the four members of the second wave of migration share relations with West Africa and the Lakes Plateau while members of the first and older wave in Siwa have relations with East Africa. The first wave suggests a Middle Pleistocene relation of the Egyptian Nile with East Africa via central Sudan while the second wave suggests that connection with West Africa came about in the Late Pleistocene/Early to Mid-Holocene wet phase (and probably as late as the Neolithic wet phase). In this period a series of higher Nile floods are synchronous with

122

greater Wadi activity in Egypt (Butzer & Hansen 1965 and 1966). Before the Neolithic wet phase, large isolated expanses of marshland patches subsequent to sudden outbursts of floods appeared in northern Sudan (de Heinzelin & Paepe 1965). These conditions must have favoured speciation of aquatic oligochaetes in Nubia where conditions are now extremely unfavourable.

Although biotite (from Bahr el-Ghazal) and pyroxene (from the Blue Nile) appear almost together in Egyptian sediments dating from about 50,000 years ago (Butzer & Hansen 1968), no faunal element from the Ethiopian highlands appears in the Delta with the faunule having a West African affinity. This can be explained by the exceedingly torrential nature of the Blue Nile flood (which destroys[1] even plankton, Talling & Rzóska 1967) in contrast with the slow White Nile debit. Neither do any aquatic oligochaetes originating from the Ethiopian highlands appear in the Khartoum region nor is any species of *Alma* found along the Blue Nile. In addition to strong floods, banks of the Blue Nile are covered by large gravels unsuitable for *Alma*.

Although the Nubia Nile acted as a focus of distribution in periods of favourable rainfall, it is now characterized by a paucity of the oligochaete fauna. Search at Aswan and the Dongola Reach did not reveal the presence of *Alma*. There is only one record of *A. nilotica* from Wadi Halfa (Brinkhurst & Jamieson 1971, p. 787). The same species was found at the Sabaluka Gorge (60 km north of Khartoum, in mud, new record). Its presence at Wadi Halfa must be regarded as a vestige of an earlier uniform distribution in the Nubia Nile. This species needs slightly sloping mud banks to obtain food and shelter. Such banks are very rare in Nubia. The geomorphology of Nubia adds to the harshness of the physical environment by features mostly of Recent origin which must have obliterated the oligochaete fauna of earlier periods. Some of these features are the cataracts, scarcity of alluvial soil[2], sand encroachments, the deeply cut valley, strong current and absence of backwaters, swamps and ponds. The flora of Nubia shows a similar phenomenon of obliterated existence. Upper Egypt shares 185 riverain plant species with central Sudan of which only 124 exist along the Nubia Nile. The 61 missing species must have existed formerly in Nubia but disappeared under the recent harsh conditions of excessive heat and drought. Human manipulation could also be an important factor.

A number of exotic species have succeeded in establishing themselves in the Nile basin. It is believed that *Eisenia rosea* of Siwa, which is related to the Sardinian population of the species could have entered the Oasis from the west during the same last Pleistocene major pluvial which helped the Ethiopian species now inhabiting the Oasis in crossing the

[1] Ed. It does not 'destroy', it displaces and thins out.
[2] Ed. This situation is now changed, see ch. 19.

Mediterranean coastal stretch starting from the Delta. Another population of *Eisenia rosea*, together with *Allolobophora jassyensis* var. *orientalis* crossed Sinai from Palestine to the Delta in the Late Pleistocene/Early to Mid-Holocene wet phase. *Allolobophora caliginosa*, *trapezoides*, *Branchiura* and *Pheretima* entered by virtue of the great horticultural activities of the 19th and 20th centuries (Khalaf & Ghabbour 1968a and b). The former is present also in Siwa Oasis. In the Sudan, only *Pheretima elongata* could establish itself. It was found in Dongola, Khartoum, Wad Medani, Singa, Malakal and Gebel Marra (new records). Its entry could have possibly taken place from the east through modern trade activities.

REFERENCES

Ball, J. 1939. Contributions to the geography of Egypt. Survey & Mines Dept., Min. Finance, Egypt: 308 pp.

Beddard, F. E. 1894. On two genera, comprising three new species of earthworms, from Western Tropical Africa. Proc. Zool. Soc. Lond.: 379–390.

Brinkhurst, R. O. 1966. A contribution toward a revision of the aquatic Oligochaeta of Africa. Zool. Afr. 2: 131–166.

Brinkhurst, R. O. & Jamieson, B. G. M. 1971. Aquatic Oligochaeta of the world. Oliver & Boyd: 860 pp.

Butzer, K. W. & Hansen, C. L. 1965. On Pleistocene evolution of the Nile Valley in southern Egypt. Canad. Geog. 9: 74–83.

Butzer, K. W. & Hansen, C. L. 1966. Upper Pleistocene stratigraphy in southern Egypt. In: W. W. Bishop & J. D. Clark, eds. Background to evolution in Africa. Univ. Chicago Press: 329–356.

Butzer, K. W. & Hansen, C. L. 1968. Desert and river in Nubia: geomorphology and prehistory environments at the Aswan reservoir. Univ. Wisconsin Press: 553 pp.

Furon, R. 1967. De L'importance de la géologie et de la paléogéographie en biogéographie. C. R. Séan. Soc. Biogéog. 380–382: 1–6.

Gavrilov, K. 1967. Los oligoquetos megadrilos de la colleccion del Musée Royal de L'Afrique Centrale, Tervuren. I.-Ocnerodrilidae. Ann. Mus. Roy. Afr. Cent. Tervuren Sér. Sci. zool. 161: 153 pp.

de Heinzelin, J. 1964. Palaeoecological conditions of the Lake Albert-Lake Edward Rift. In: C. Howell & F. Bourlière, eds. African ecology and human evolution, London: 285–303.

de Heinzelin, J. 1966. Pleistocene sediments and events in Sudanese Nubia. In: W. W. Bishop & J. D. Clark, eds. Background to evolution in Africa. Univ. Chicago Press: 313–328.

de Heinzelin, J. & Paepe R. 1965. The geological history of the Nile Valley in Sudanese Nubia. Preliminary results. In: F. Wendorf, Contributions to the prehistory of Nubia, Fort Burgwin Res. Cent. and S. Methodist Univ. Press, Dallas Texas 2: 29–56.

Jamieson, B. G. M. 1957. Some species of *Pygmaeodrilus* (Oligochaeta) from East Africa. Ann. Mag. Nat. Hist. (12) 10: 449–470.

Jamieson, B. G. M. 1969. A new Egyptian species of *Chuniodrilus* (Eudrilidae, Oligochaeta) with observations on internal fertilization and parallelism with the genus *Stuhlmannia*. J. nat. Hist. 3: 41–51.

Jamieson, B. G. M. & Ghabbour, S. I. 1969. The genus *Alma* (Microchaetidae: Oligochaeta) in Egypt and the Sudan. J. nat. Hist. 3: 471–484.

Khalaf El-Duweini, A. & Ghabbour, S. I. 1967. Records of Oligochaeta in Egypt. Pedobiologia 7: 135–141.

Khalaf El-Duweini, A. & Ghabbour, S. I. 1968a. The geographical speciation of northeast African oligochaetes Pedobiologia 7: 371–374.

Khalaf El-Duweini, A. & Ghabbour, S. I. 1968b. The zoogeography of oligochaetes in north-east Africa. Zool. Jb. Syst. 95: 189–212.

Maley, J. 1969. Le Nil: données nouvelles et essai de synthèse de son histoire géologique. Bull. Liais. ASEQUA, Dakar 21: 40–48.

Maley, J. 1970. Introduction à la géologie des environs de la deuxième cataracte du Nil au Soudan. In: J. Vercoutter, ed. 'Mirgissa I', Mission Archéol. Fran. au Soudan, Libr. Orient., Paris: 122–157, 2 c.h.t.

McBurney, C. B. M. 1960. The stone age of northern Africa. Pelican Books: 288 pp.

Murray, G. W. 1949. Desiccation in Egypt. Bull. Soc. Geog. Égypte 23: 19–34.

Pfannenstiel, M. 1953. Das Quartär der Levante. II. Die Entstehung der ägyptischen Oasendepressionen. Ak. Wiss. Lit. Math-Nat. Kl., Mainz: 335–441.

Talling, J. F. & Rzóska, J. 1967. The development of plankton in relation to hydrological regime in the Blue Nile. J. Ecol. 55: 637–662.

7. FISH FAUNA OF THE NILE

by

P. H. GREENWOOD

7a. *Zoogeography and History*

Taken in its entirety, that is by including the fishes of Lakes Victoria, Kioga, Edward, George, Albert and Tana, the ichthyofauna of the Nile drainage basin is the second largest in Africa. (The Congo region, with 669 species according to Poll (1973), is the largest). This 'extended Nile' fauna comprises some 320 species (in 60 genera), of which approximately 62 percent are endemic.

However, although Lakes Victoria (with Kioga) and Edward (with George) are connected by river with the Nile, the lake faunas are physically isolated by barriers that seemingly are impassable to fishes. The Murchison Falls isolate the fishes of Lakes Victoria and Kioga, and the Semliki rapids cut off the fishes of Lakes Edward and George from those of Lake Albert and the Nile. It is the highly endemic and taxonomically diversified faunas of these four lakes that give the extended Nile fauna its status amongst the ichthyofaunal provinces in Africa. (The provinces are those defined by Poll 1973).

If the fishes of these lakes are excluded from any consideration of the Nile's zoogeographical relationships, the Nile can in fact be looked upon as a northern extension of the Occidental province (Niger, Senegal and Volta river systems), and in particular an extension of the Niger.

Clearly the major evolutionary foci are in lakes isolated from other water bodies by barriers impassable to fishes. For example, Lake Albert, a major lake by any physical parameters, is not isolated in this way. Its fauna, apart from about six endemic cichlid species, one endemic cyprinid, and one centropomid species (*Lates macrophthalmus*; see Holden 1967) is otherwise shared with the Nile, although the river does have a few species (both endemic and non-endemic) which apparently do not colonise Lake Albert (Greenwood 1966).

Lakes Rudolf and Tana must also be considered in relation to the Nile. Although Rudolf is no longer in contact with the Nile there is strong evidence for its having had a direct river connection in the past (see summary in Greenwood 1974b). Unlike the other isolated lakes, Rudolf does not have a highly endemic or differentiated fish fauna (Worthington & Ricardo, 1936). In these respects its fish fauna is about equal to, or slightly less developed than that of Lake Albert (Greenwood 1974b and below; also personal observations and communications *in litt.* from Mr and Mrs A. J. Hopson). Several Nile species well represented in Lake

Albert (especially members of the family Mormyridae) are, however, absent from Lake Rudolf. The absence of these species may be attributable, in part, to the lake's peculiar chemistry (especially its high alkalinity, see Talling & Talling 1965), while the absence of endemic species may be correlated with the relatively short time Lake Rudolf has been in isolation (Fryer & Iles 1972; Greenwood 1974b).

Unfortunately, the fishes of Lake Tana are not at all well studied. The little information available shows the lake to have an extremely truncated Nile type fauna of some eight or nine species, one or two of which are endemic. At least one *Barbus* species *(B. intermedius)* is of limited distribution outside Lake Tana (Banister 1973), and the lake also contains specics that, although not endemic to it, do not occur in the main Nile or its associated lakes (as for example, *Varichorinus beso* which does enter the Blue Nile). An outstanding feature of Lake Tana's ichthyofauna is the cobitid species *Noemacheilus abyssinicus*, the only known representative of this Euro-Asiatic genus to occur in Africa.

Returning now to the zoogeographical relationships of the Nile itself, and in particular its affinities with the western rivers. Of the 115 species recorded in the Nile, some 26 (or 23 per cent) can be considered endemic. (The concept of endemism here being stretched a little to allow for the fact that some of these species occur also in the Nile basin lakes; none, however, extends beyond this region).

Seventy-four of the 115 species in the Nile also occur in the Niger, and 22 of these are found in the Zaire (formerly Congo) river as well. The majority of species common to the Nile and Niger are, of course, found in the other major west African rivers (the Volta, Senegal and Gambia) that go to make up the Occidental ichthyofaunal province. Only some ten species are restricted to the Nile, Niger and Chad basins. As might be expected, the Niger has its share of endemics, about 24 out of a total of between 120 and 130 species. (Data derived mainly from Daget 1954; Hopson 1967; Blache 1964 and Kähsbauer 1962). Thus, in their basic similarities and in the size, diversity and endemism of their fish faunas, the Nile and Niger are closely comparable. To emphasise this similarity, it may be noted that the closest relatives of the endemic species in each river are to be found in the other river, thus strongly suggesting their derivation from a common ancestor.

Lake Chad, as might be expected from its geographical position, has an essentially Nilotic fauna with an admixture of species (at least 17 out of a total of 87) that otherwise are Niger-Occidental elements. Eight primarily Nilotic species extend no further westwards than the Chad basin. (Data from Hopson 1967 and Blache 1964). Presumably Lake Chad's fish fauna is composed partly of a relict one (that is, species with a Nile, Occidental and even Congoan distribution) and partly of later immigrants that had evolved in the Nile basin after direct connection between the Nile and Occidental regions had been lost. In other words the Lake Chad

128

area now acts as a differential filter between the Nile and Occidental provinces.

At least two Nile species are found in the Webi Shebeli system of Ethiopia and Somalia. Since the range of neither species (the cyprinid *Chelaethiops bibi*, and the mochokid *Synodontis frontosus*) extends beyond the Nile-Chad basins it seems likely that both are invaders of the Webi Shebeli rather than relict species (as would seem to be the status of *Synodontis schall* the other non-endemic mochokid in that system).

Relict status seems certain for the five Nile species (*Protopterus aethiopicus*, *Polypterus bichir*, *P. senegalus*, *Ichthyborus besse* and *Sarotherodon* (= *Tilapia*) *niloticus*) occurring in the Lualaba province (Poll, 1973), a major affluent system of the Zaire river. Except for *Protopterus aethiopicus*, all these species are also found in the Occidental province, while *P. aethiopicus* extends its range well into the Zaire system (see fig. 5 in Poll *op. cit.*). Seen in this light these five species can no longer be used, as they are by Poll *(op. cit.)*, to argue that the Lualaba river was once a tributary of the Nile.

A consideration of the Nile river ichthyofauna as a whole, and especially the inferences to be drawn from its close affinities at the species level with the fauna of the Occidental province, leads to one conclusion. Namely, that the Nile fishes are a persistent segment (albeit a large one) of an ancient African fish fauna that was once widely distributed north of latitude 10 °S.

Unfortunately there is no way of determining the time span over which this faunal assemblage developed. The available fossil evidence gives little help in this respect (see Greenwood 1974c for summary), mainly because few fossil remains are identifiable to the species level. Generic identification can be of great value when studying certain aspects of a particular zoogeographical history (for example the history of a water body like Lake Edward with its characteristic fauna, see Greenwood 1959; Greenwood & Howes 1975, and below). On a broader scale, however, genera are of little value because most have a pan-African distribution; it is largely the species that give an ichthyofaunal province its diagnostic features.

Undoubtedly the most interesting fossils so far discovered in the Nile valley are some from Pliocene deposits near Wadi Natrun (Greenwood 1972). These are the remains of an extinct characin species whose affinities are more with South American representatives of this family (eg. *Myletes*) than with the African genera. Recently this extinct species has been identified from early Pleistocene deposits near Lake Albert, and a related species has been found in Miocene deposits between Lake Albert and Lake Edward (Greenwood & Howes 1975).

At least some Nilotic-Occidental genera and species (*Lates niloticus*, *Clarias* sp., *Heterobranchus* sp., *Chrysichthys furcatus*, *Clarotes laticeps*, *Arius gigas*, *Auchenoglanis occidentalis*, *Bagrus docmac*, *Synodontis schall*, *S. nebulosus*,

129

S. ocellifer, and a species of *Tilapia* or *Sarotherodon*) were widely distributed in the Sahara region during Upper or Neo-Pleistocene times (see Green-wood 1974c for summary). Even today there exists in the Ennedi Plateau region of the Chad Republic a small cyprinid, *Barbus apleurogramma* which otherwise occurs only in the Nile basin (Lake Victoria region) and in Tanzania (Greenwood 1966), as well as eight Nilo-Occidental species (cyprinids, clariids, cyprinodontids and cichlids; see Daget 1959).

In conclusion, some thought should be given to the marked discrepancy in the number of species (and the degree of endemicity) between the fishes of the Zaire and Nile rivers (endemic species from the Nile basin lakes being excluded in this comparison).

In the Nile there are 115 species of which 26 (*ie* about 23 per cent) are endemic. The Zaire river has 669 species of which 548 (*ca* 82 per cent) are endemics (Poll 1973). This great disparity in numbers and degree of endemism is the more remarkable when one considers that, in all proba-bility, both faunas were derived from the same ancestral species and that both have been evolving over the same time period. The known differences in ecological diversity (and hence of evolutionary opportunity) between the two rivers is alone unlikely to account for the greater number of species in the Congo (see Lowe-McConnell 1969), although this must be an important contributory factor (perhaps more so in the past than now, see below).

Recently, Roberts (1972) has developed with respect to the Zaire an idea first put forward by Myers (1960) to explain the ichthyofaunal diversity of the Amazon. Essentially, this hypothesis stems from the observation that many tropical lake faunas are more diverse and differen-tiated than are those of neighbouring rivers, a point well illustrated in Africa. Roberts suggested, therefore, that in its past history the Zaire basin was subdivided into a number of lakes, during the develop-ment of which there had been a concomitant diversification of the fishes in each lake. The fauna of the present day river represents an analgam of these once lacustrine faunas. If this was the Zaire's history one must assume that few lake-like water bodies were involved in the past history of the Nile, and indeed this does seem to have been so (Berry & Whiteman 1968).

Although the Myers-Roberts hypothesis is an attractive one, a great deal more information is needed on the phyletic interrelationships of the Zaire fishes, as well as further information of the river's physiographical history, before it can be tested adequately. One outstanding point con-cerns the relative proportions of cichlid to non-cichlid fishes in the river. Cichlids dominate most African lake faunas, but non-cichlids predominate in the rivers (see Lowe-McConnell 1969; Fryer & Iles 1972; Greenwood 1974a). Both the Zaire and the Nile are typical in this respect. If the Zaire went through a 'multiple lake' stage, one can reasonably assume that the cichlids evolved therein might be less successful under fluviatile

130

conditions. Hence cichlids would, as is the actual case, be poorly represented in the resulting river. But, the multiplicity of non-cichlid species in the present day Zaire is much greater than one would expect, on the basis of existing lake faunas, to have evolved in the hypothetical lakes of the proto-Zaire (see Table I in Lowe-McConnell 1969). Clearly we are still far from understanding why the Zaire ichthyofauna should be so much more diversified than that of the Nile, nor do we yet have a great deal of information on which to hypothesise about the early history of the Nile fish fauna.

7b. Nile Fishes, General

In the previous chapter reference was made to the 'extended Nile' ichthyofauna. The term was coined to cover the fish faunas of the main river and its affluents, and those of the lakes lying within the Nile drainage basin. The diversity of fishes in some of these lakes, the high level of their endemicity, and the evolutionary problems posed by such phenomena, are now well documented and discussed (Fryer & Iles 1972; Greenwood 1966, 1973, 1974a and b). The fascination these problems hold for biologists, coupled with the economic importance of many lake fishes, have resulted in rather more research being done on the fishes of the Nilotic lakes than on the species inhabiting the Nile itself.

Broadly speaking, Nile basin lakes can be divided into two types. First, there are those in which the predominant faunal elements are fishes of the family Cichlidae, and in which there is a high level of endemicity in both cichlid and non-cichlid fishes (eg. Lakes Victoria, Kioga, Edward and George). In the second category are those lakes whose faunas show a low level of endemicity, and in which cichlid species do not predominate (eg. Lakes Albert, Rudolf and Tana); see Table.

Lakes in the second category have fish faunas that are essentially like the fauna of the main Nile but are slightly depauperate in terms of species numbers, and usually have a small number of endemic species, generally cichlids of the genus *Hapiochromis*, or small cyprinids (see Greenwood 1966 and 1974b; Worthington & Ricardo 1936).

Lake Rudolf, now completely isolated from the Nile has, on the whole, a fauna very similar to that of Lake Albert, including many species shared by both lakes and by the Nile itself. Some of the endemic cichlid species and the one endemic *Lates* species in each lake show marked parallel evolution in their adaptations to deep-water habitats (Greenwood 1974b, and unpublished observations). A few endemic non-cichlid subspecies have been described from Lake Rudolf (Worthington & Ricardo 1936), a reflection of the lake's geographical isolation and of the relatively short time over which that condition has been effective.

Lake Tana, as far as can be told from its poorly studied fauna, should be considered merely as part of the Blue Nile (see below, also Banister

Table 1. Analysis of the fishes in the Nile basin lakes. Lakes isolated from the Nile are marked thus *. See text, page 131.

CICHLIDAE

Lake	Number of species		*Haplochromis* spp.	Endemic *Haploch-romis* spp.	Number of genera	
	Total	Endemic			Total	Endemic
*Victoria	ca 150–170	All but 3	All but 8	All but 1	8	4
*Edward-George	ca 35– 40	All but 5	All but 4	All but 1	4	0
Albert	10	4	6	4	2	0
*Rudolf	8	3	3	3	3	0

OTHER FAMILIES (ie. non-cichlidae)

Lake	Number of families	Number of species		Number of genera	
		Total	Endemic	Total	Endemic
*Victoria	11	38	16	20	1
*Edward-George	7	17	2	10	0
Albert	13	36	3	21	0
*Rudolf	14	32	5	22	0

1973). However, the occurrence in Tana of an exotic faunal element, the cobitid *Noemacheilus abyssinicus* gives this small lake a particular character, and one that remphasises the need for further detailed studies on its fishes.

Lakes in the first category present a very different picture from those already described. Each has an ichthyofauna that must be thought of as a highly derivative Nile type. Indeed, their fish faunas have differentiated so far from the Nile type that there is every justification for considering these lakes as making up a distinct ichthyofaunal province, the Edwardo-Victorian[1]. Of the two lakes, Edward has the closer ties with the Nile because of the relatively higher proportion of Nile species in its fauna (Greenwood 1966), including two Nilotic species of *Sarotherodon*. (In Lake Victoria the naturally occurring *Sarotherodon* species are endemic).

Both Lakes Victoria and Edward have fluviatile connections with the Nile (Bahr el Jebel), Lake Victoria directly *via* the Victoria Nile, and Lake Edward rather indirectly through the Semliki river and Lake Albert. But, and this is important, the fishes are confined to the lakes by impassable barriers (see above).

Characteristic features of the Edwardo-Victorian fauna, apart from the taxonomic and ecological predominance of cichlid species, are the absence of twenty-five Nile genera (including *Lates*, *Polypterus*, *Hydrocynus*, *Hyperopisus* and *Mormyrops* among the larger fishes), and the presence of numerous endemic noncichlid species (at least as many, for the lakes combined, as occur in the Nile). Furthermore, there is evidence to suggest that the dominant cichlids of the region (species of the genus *Haplochromis*) are phylogenetically distinct from those of the Nile, Lake Albert and Lake Rudolf (Greenwood 1974b).

Historically, Lakes Victoria and Edward should perhaps be thought of as late-comers to the Nile drainage system (see Greenwood 1973 and 1974a), probably not contributing to the system until Pleistocene times (See ch. 3). Recent geological research suggests that Lake Albert should also be considered in this category (Berry & Whiteman 1968). That all three lakes have, or had a Nile-type ichthyofauna can be explained on the basis of that faunal type being an archaic one whose distribution over a large part of Africa preceded the formation of the lakes.

The excellent Pliocene to late Pleistocene fossil record for Lake Edward (Greenwood 1959; Greenwood & Howes 1975) reveals that during this period the fish fauna of the lake was more Nile-like than it is today, and included several of the genera now absent. Regrettably the fossil record for Lake Victoria is virtually non-existent save for indications of certain Nile-type elements (now absent) being present in the area before the modern lake was formed (Greenwood 1951; 1974a).

---

[1] For the purposes of this discussion only Lake Kioga can be taken as part of Lake Victoria, and Lake George as part of Lake Edward.

Nile fishes is compared with that in either Lake Victoria or Lake Edward. Seasonal breeding would seem to be the mode in the Nile, continuous breeding in the lakes. Closer examination shows that it is, in fact, the non-cichlids which breed seasonally, whether they are fluviatile or lacustrine. The Cichlidae are the continuous breeders, no matter in what environment the species may live (Greenwood 1966 and 1974a; Sandon 1950; and notes, by various authors, published in the Annual Reports for the Hydrobiological Research Unit of Khartoum University). However, even the cichlids show an increase in spawning activity during the rainy and correlated flood seasons.

In both these respects the Nile is more like Lake Albert than it is either of the cichlid-dominated, isolated lakes. When more is known about the biology of fishes in the non-cichlid dominated (although isolated) Lake Rudolf, it will probably be found to have Nilotic characteristics in these respects also.

The Blue Nile ichthyofauna does not seem to be completely representative of the main Nile species, although the majority of genera are represented. To date 46 species have been recorded from the Blue Nile, but with most of the species apparently confined to the river downstream of Roseires (Abu Gideiri 1967). Almost certainly this impression (and the number of species) will be altered when the middle and upper reaches of the Blue Nile are adequately sampled. Nevertheless, it is still surprising that so few species are known from the lower reaches of the river. In some contrast to this situation there are two species (*Barbus intermedius* and *Varichorinus beso*) restricted to the upper part of the river (Banister 1973; also above). Presumably ecological barriers can be invoked to explain these distributional peculiarities.

Another major affluent of the White Nile, (B. el Gebel) the Aswa river, has a most intriguing fauna, one that is possibly unique amongst the Nile's several tributaries. The Aswa arises on high ground in western Karamoja (Uganda), and joins the Bahr el Gebel about 30 miles north of Nimule. Although the source of the Aswa is close to the Lake Kioga drainage basin, there is, at present, no interconnection between the two systems. That some linkage existed in the past seems likely from the fact that the fishes of the Aswa, even in regions near its source, are an admixture of Nile species not otherwise occurring in the Victoria-Kioga system, of species endemic to that system, and of species with a wider geographical range but not occurring in the Nile or Lake Albert (Greenwood 1963[1]).

The high proportion (14:9) of non-Nile to Nile species suggests that in former times the river was part of the Victoria drainage system, and that

[1] The presumed endemic Aswa *Haplochromis* species *(H. cancellus)* is now known to be based on a specimen of the Lake Malawi species *H. taeniolatus* accidentally packed with the Aswa collection before its dispatch from Uganda.

136

the Nile species were later invaders which gained access after the river's incorporation into the Nile drainage.

The large number of species shared by the Nile and rivers in western Africa has been noted already. There is a corresponding similarity in the roles played by the fishes in the total ecology of the rivers. Usually the same species occupies the same niche in different rivers, and even when there is geographical replacement of species their ecological roles remain similar.

That the Zaire river has five times as many species as the Nile (Poll 1973), and that these fishes exhibit far greater ecological diversity, can probably be explained by the different histories of the lake basins. Also to be taken into account are the greater variety of habitats provided by the Zaire (including several lake-like areas and torrential rapids). Climatic factors too may exert their influence, the equatorial Zaire showing much less seasonal variation than the Nile. Nowhere does the Zaire traverse desertic areas; indeed in many places it flows through dense forest (see Lowe-McConnell 1969, for further discussion of these points).

There are certain interesting differences between the ichthyofaunas of the Nile on the one hand and those of the Zaire and the rivers in west Africa on the other hand. Particularly noticeable are the absence from the Nile of some families and genera which are well-represented in the Congo and Occidental regions.

A good example is provided by the herring-like fishes of the suborders Clupeoidei and Denticipitoidei. There are several strictly freshwater representatives of the Clupeoidei in the Zaire and Niger rivers, but none in the Nile or any part of its drainage basin. The Denticipitoidei, represented by the single genus and species *Denticeps clupeoides* (Clausen 1959), is known from two localities in the Niger system (Clausen, *op. cit.* Gras, 1961, and Greenwood, 1965). *Denticeps* is a tiny but most distinctive fish characterized by a number of primitive and specialized anatomical features (Greenwood 1968). A fossil species *Palaeodenticeps tanganyikae* (Greenwood 1960), very like the extant species, is known from Palaeogene (?Miocene) deposits to the southwest of Lake Victoria. This record would indicate an earlier distribution of the denticipitoids in an area that could have had connections with the Nile system.

Rivers in the Congo and Occidental ichthyofaunal provinces (see Poll 1973, and above) have more representatives of the Osteoglossoidei and Notopteroidei than does the Nile. Only *Heterotis niloticus* and *Xenomystus nigri* occur in the latter river, but both these species and two others, *Pantodon buchholzi* and *Papyrocranus afer*, are widely distributed in the Congo and Occidental regions. These species, all members of the Osteoglossomorpha, an extremely primitive group of bony fishes (Greenwood *et al.* 1966), must be considered together with the lungfishes and

137

the Bichirs (Polypteridae) as some of the most archaic components in the African ichthyofauna.

Another primitive group, the Gonorynchiformes (see Rosen & Greenwood 1970), is also better represented in the west (especially in the Zaire river) than in the Nile (see generic distribution map in Poll, 1973). A single genus and species, *Cromeria nilotica*, occurs in the Nile (and Niger), whereas three genera and several species are found in the Zaire (but, it should be noted, only two genera and species in the Niger). No gonorynchiform is found in any lake of the Nile system.

There are other but less spectacular cases of absenteeism in the Nile. For example, the genus *Mastacembelus* (Mastacembelidae) well represented by numerous species (including an eyeless form) in the Zaire river, are entirely absent from the Nile basin except in the Aswa river and Lake Victoria. Mastacembelids, incidentally, are poorly represented in the Occidental region (Daget 1954).

At a more subtle level, it seems that the majority of *Barbus* species inhabiting the Congo and Occidental provinces belong to a different phyletic lineage from those species in the Nile basin, including the lakes (Greenwood, unpublished observations).

For the moment these inter-provincial differences are inexplicable. More refined taxonomic revisions, a better fossil record, and further geomorphological research are needed before any worthwhile hypothesis can be formulated.

Moving southwards from the Congo and Occidental regions it need only be noted that the Nile river has a somewhat more diversified ichthyofauna than the Zambesi province (*ca* 115 *cf* 101 species), but a less diverse one than that of the Angolan province (*ca* 270 species; Poll 1967). There is, as would be expected, a rather distant relationship at the species level between these latter faunas and that of the Nile. Only six Nile river species[1] occur in the Zambesi river, and the same six are found in the Zaire; all but *Aplocheilichthys hutereaui* also occur in the rivers of Angola, and the range of two species, *Hydrocynus vittatus* and *Micralestes acutidens*, even extends to the Limpopo river (total number of species *ca* 60).

Below these latitudes no Nile river species are found, although at least one *Barbus* species from the Lake Victoria basin (and Ethiopia) extends to South Africa (Greenwood 1962).

The affinities of the Nile ichthyofauna with the as yet poorly studied fishes of the Oriental region (Ethiopia, Somalia together with eastern Kenya, Tanzania and Mozambique) are briefly considered above.

What we have considered so far is the product of past evolutionary changes, both biological and physiographical. What of the future?

In recent years human activities have begun to have a marked influence

---

[1] Viz. *Hydrocynus vittatus, Micralestes acutidens, Schilbe mystus, Heterobranchus longifilis, Malapterurus electricus, Aplocheilichthys hutereaui.*

138

on the Nile ichthyofauna. During the last decade several species, including the Nile Perch, *Lates niloticus,* have been introduced into the Lake Victoria basin (Gee 1969), the water hyacinth *(Eichhornia crassipes)* has gained access to the Nile, dams have been built in many places on the main river and its tributaries, and there has been a great increase in the intensity of fishing pressure, at least on the lakes (Fryer 1972). Already some ecological effects of this interference have been observed (Greenwood 1957; Bishai 1959; Hammerton 1972). Whatever the future results of these changes may be (and it is difficult to be optimistic), new and fast-acting evolutionary factors have been brought to bear on the Nile basin fish fauna. The next century could well see changes as profound as those that have taken place in the last two million years.

REFERENCES

Abu Gideiri, Y. B. 1967. Fishes of the Blue Nile between Khartoum and Roscires. Revue Zool. Bot. Afr., 76: 345–348.
Banister, K. E. 1973. A revision of the large *Barbus* of east and central Africa. Bull. Br. Mus. nat. Hist. (Zool.), 26: 1–148.
Berry, L. & Whiteman, A. J. 1968. The Nile in the Sudan. Geogr. J. 134: 1–37.
Bishai, H. M. 1959. The effect of *Eichhornia crassipes* on larval and young fish. 6th Annual Rep. Hydrobiol. Res. Unit, Khartoum Univ.: 19–22.
Blache, J. 1964. Les poissons du Tchad et du bassin adjacent du Mayo Kebbi. O.R.S.-T.O.M. Paris.
Boulenger, G. A. 1907. The fishes of the Nile. Hugh Rees, London.
Clausen, H. S. 1959. Denticipitidae, a new family of primitive isospondylous teleosts from West African freshwater. Vidensk. Meddr. dansk. naturh. Foren., 121: 141–151.
Corbet, P. S. 1961. The food of non-cichlid fishes in the Lake Victoria basin, with remarks on their evolution and adaptation to lacustrine conditions. Proc. zool. Soc. Lond., 136: 1–101.
Daget, J. 1954. Les poissons du Niger supérieur. Mem. Inst. Fr. Afr. noire, 36: 1–391.
Daget, J. 1959. Note sur les poissons du Borkou-Ennedi-Tibesti. Trav. Inst. Rech. Sahariennes Univ. d'Alger., 18: 173–181.
Fryer, G. 1972. Conservation of the great lakes of east Africa: a lesson and a warning. Biol. Conserv., 4: 256–262.
Fryer, G. & Iles, T. D. 1972. The cichlid fishes of the Great Lakes of Africa. Their biology and evolution. Oliver and Boyd, Edinburgh.
Gee, J. M. 1969. A comparison of certain aspects of the biology of *Lates niloticus* (Linne) in endemic and introduced environments in east Africa. In: Man-Made Lakes (Ed. L. E. Obeng): 251–260. Ghana Universities Press, Accra.
Gras, R. 1961. Contribution à l'étude des poissons du Bas-Dahomey. Description de quatre espèces nouvelles. Bull. Mus. Hist. nat. Paris (2), 32: 401–410.
Greenwood, P. H. 1951. Fish remains from Miocene deposits of Rusinga Island and Kavirondo Province, Kenya. Ann. Mag. nat. Hist (12), 4: 1192–1201.

139

Greenwood, P. H. 1957. Two new records of non-cichlid fishes from Lake Victoria and the Victoria Nile. In: 2nd Symposium on African hydrobiology and inland fisheries. Cons. sci. Afr. Publ. No. 25: 79.

Greenwood, P. H. 1959. Quaternary fish fossils. Explor. Parc. Natn. Albert Miss. J. de Heinzelin de Braucourt, 4: 1–80.

Greenwood, P. H. 1960. Fossil denticipitid fishes from east Africa. Bull. Br. Mus. nat. Hist. (Geol.), 5: 1–11.

Greenwood, P. H. 1962. A revision of certain *Barbus* (Pisces, Cyprinidae) from east, central and south Africa. Bull. Br. Mus. nat. Hist. (Zool.), 8: 151–208.

Greenwood, P. H. 1963. A collection of fishes from the Aswa river drainage system, Uganda. Proc. zool. Soc. Lond., 140: 61–74.

Greenwood, P. H. 1965. On the status of *Acanthothrissa* Gras, 1961 (Pisces, Clupeidae), Ann. Mag. nat. Hist. (13), 7: 337–338.

Greenwood, P. H. 1966. The fishes of Uganda (2nd Edn.). Uganda Society, Kampala.

Greenwood, P. H. 1971. On the cichlid fish *Haplochromis wingatii* (Blgr.), and a new species from the Nile and Lake Albert. Revue Zool. Bot. afr., 84: 344–365.

Greenwood, P. H. 1972. New fish fossils from the Pliocene of Wadi Natrun, Egypt. J. Zool. Lond., 168: 503–519.

Greenwood, P. H. 1973. A revision of the *Haplochromis* and related species (Pisces, Cichlidae) from Lake George, Uganda. Bull. Br. Mus. nat. Hist. (Zool.), 25: 139–242.

Greenwood, P. H. 1974a. The cichlid fishes of Lake Victoria, east Africa: the biology and evolution of a species flock. Bull. Br. Mus. nat. Hist. (Zool.), Suppl. 6: 1–134.

Greenwood, P. H. 1974b. The *Haplochromis* species (Pisces, Cichlidae) of Lake Rudolf, east Africa. Bull. Br. Mus. nat. Hist. (Zool.), 27: 139–165.

Greenwood, P. H. 1974c. Review of Cenozoic freshwater fish faunas in Africa. Ann. Geol. Surv. Egypt. 6: 211–232.

Greenwood, P. H. & Howes, G. J. 1975. Neogene fishes from the Lake Albert-Lake Edward rift (Zaire). Bull. Br. Mus. nat. Hist. (Geol.), 25 (in the press).

Greenwood, P. H., Rosen, D. E., Weitzman, S. H. & Myers, G. S. 1966. Phyletic studies of teleostean fishes with a provisional classification of living forms. Bull. Am. Mus. nat. Hist., 131: 339–456.

Hammerton, D. 1972. The Nile river – a case history. In: River ecology and man (Eds. R. T. Oglesby, C. A. Carlson & J. A. McCann): 171–214. Academic Press, London.

Holden, M. J. 1967. The systematics of the genus *Lates* (Teleostei: Centropomidae) in Lake Albert, east Africa. J. Zool., Lond., 151: 329–342.

Hopson, A. J. 1967. The fisheries of Lake Chad. In: Fish and fisheries of Northern Nigeria. (W. Reed, J. Burchard, A. J. Hopson, J. Jenness & I. Yaro): 189–200. Ministry of Agriculture, Northern Nigeria.

Kähsbauer, P. 1962. Beitrag zur Kenntnis der Fischfauna von Nigeria. Annln. naturh. Mus. Wien, 65: 139–165.

Lowe-McConnell, R. H. 1969. Speciation in tropical freshwater fishes. Biol. J. Linn. Soc. 1: 51–75.

Myers, G. S. 1960. The endemic fish fauna of Lake Lanao, and the evolution of higher taxonomic categories. Evolution, 14: 323–333.

Poll, M. 1967. Contribution à la faune ichthyologique de l'Angola. Publições cult. Co. Diam. Angola, 75: 1–381.

Poll, M. 1973. Nombre et distribution géographique des poissons d'eau douce africains. Bull. Mus. Hist. nat. Paris (3), no. 150: 113–128.

Roberts, T. R. 1972. Ecology of fishes in the Amazon and Congo basins. Bull. Mus. comp. Zool. Harv., 143: 117–147.

Rosen, D. E. & Greenwood, P. H. 1970. Origin of the Weberian apparatus and the relationships of the ostariophysan and gonorynchiform fishes. Am. Mus. Novit., 2428: 1–49.

Sandon, H. 1950. An illustrated guide to the freshwater fishes of the Sudan. Mc-Corquodale and Co. London (for Sudan Notes and Records).

Sandon, H. & Tayib, A. 1953. The food of some common Nile fish. Sudan Notes Rec., 34: 205–229.

Talling, J. F. & Talling, I. B. 1965. The chemical composition of African lake waters. Int. Revue ges. Hydrobiol. Hydrogr., 50: 421–463.

Worthington, E. B. & Ricardo, C. K. 1936. The fish of Lake Rudolf and Lake Baringo. J. Linn. Soc. (Zool.), 39: 353–389.

III. HYDROLOGY AND SEDIMENTS

8. NILE WATERS – HYDROLOGY PAST AND PRESENT

by

J. RZÓSKA

The Nile is hydrologically the best explored river in the world and this development arose out of necessity. An enormous amount of data exists in the volumes of 'The Nile Basin' and many other publications. Hurst (1952) has given a concise description of the Nile system, its hydrology, geography and agriculture, based on life-long service in Egypt.

Here some of the most important facts have been brought together to justify the term 'Life Artery' applied to the Nile. The Nile affected man decisively and 'fluviatile' civilizations developed along a large part of the river valley, of which Egypt was the 'crown'.

The onset of ever increasing aridity forced the growing populations from the Neolithic onwards to concentrate on the river and, as in other fluviatile settlements e.g. in Mesopotamia, skills developed for the management of water resources. In dynastic times, according to Teclaff *et al.* (1973), 'hydraulic engineering' reached a high degree of accomplishment; reclamation schemes on the left bank of the Nile were initiated during the first dynasty, a diversion canal was built from the river to the Faiyum, which acted as reservoir during the floods; dams, dykes and canals were constructed and later water-lifting machinery was invented. Dams were erected even in the eastern wadis, e.g. the Wadi Garawi, during the third or fourth dynasty; at that time these eastern desert wadis must have been still considered as water resources (Murray 1947 and Hellström 1951, quoted after Kassas 1953).

Hydrological observations in ancient times were confined to the most important event of the year – the arrival and level of the flood. Nilometers were installed to record the height of the water. Chélu (1891) names seven of these, the most important being the 'House of Inundations' presently situated in Old Cairo, one at Elephantine island near the first cataract; the gauge at Semna-Kumna existed from the Middle Kingdom onwards and, much later, the most famous Nilometer at Rhoda was constructed. This contains a record, though not continuous, from 641 A.D. onwards, the longest catalogue of river flow in the world. The Rhoda data have been analysed by Popper (1951) as to periodicity of floods and their relationship to rainfall in Ethiopia.

For the ancient Egyptians, cultivation was strictly regulated by the advent of the flood. This inundated the land, left a new layer of silt, which could be sown into with a little ploughing after the surplus of water

145

drained off into the river or downward into the existing underground water level. The growing season was in autumn and crops were harvested in March-April with the land then left fallow until the next flood. This was 'basin irrigation' practised until the 19th century. According to Hamdam (1961) 'the basin irrigation is biologically sound, making for soil conservation and regeneration. It has survived through millenia because it maintained the ecological balance of the soil ... Since it needed few canals and ditches or drains, it was likewise in tune with the principle of space economy, so imperious in such limited ... land as ... Egypt'. But this system still required careful control, which broke down in times of unrest as in late Roman times and the first Arab conquest. But the tenacity of the fellahin cannot be underrated; they cultivated their fields through the turbulent centuries of Egypt's history. The greatest changes occurred in the 19th century when the pressure of population demanded more food. Perennial irrigation was introduced providing crops during the whole year. Barrages, both temporary and permanent, and dams had to be built to provide water during the low season. By 1886 already three million acres were under perennial irrigation, in 1920 four million against 1.6 million of basin irrigation; by 1955 over six million acres were under cultivation with only 0.7 million still under basin practice. The demand for water increased and led to the building or planning of ever-larger dams and waterworks. Now, in the 20th century, the problem has become even more acute.

In fact it has become almost a political problem with the emergence of the Sudan as a consumer of water. A 'Nile Water Agreement' of 1929 was refuted by the Sudan later, when it became increasingly independent from British rule. The 'Nile Water Question' as defined by the Sudan Government (Ministry of Irrigation, Khartoum 1955) was finally settled after considerable acerbity. The Sudan agreed to the construction of the Aswan High Dam, which involved the drowning of the Wadi Halfa region and the evacuation of the entire population, against concession of a larger share of water, agreement to the building of the Roseires Dam and above all some agreement on a comprehensive plan for the 'Hydrological Unity of the Nile Valley'. We have to remember that almost every cubic metre of water in the Nile is measured. Of the ambitious plan for the whole Nile, the proposed dams at the Lake Victoria outlet, at Roseires and on the Atbara (Khasm el Girba) have been achieved; not so the Jonglei canal to bypass the Upper Nile swamps (but see ch. 12), a dam at the outlet of Lake Tana and one at Nimule on the upper White Nile, and other smaller schemes. Some of these plans refer to other countries, Uganda and Ethiopia, and will require international co-operation.

In all this the extraordinary fact remains – none of the water required by the Sudan and Egypt originates there. Both countries live on a foreign supply – Egypt is practically rainless and in the Sudan only some southern parts have a rainfall supply sufficient for seasonal crops, with long periods

146

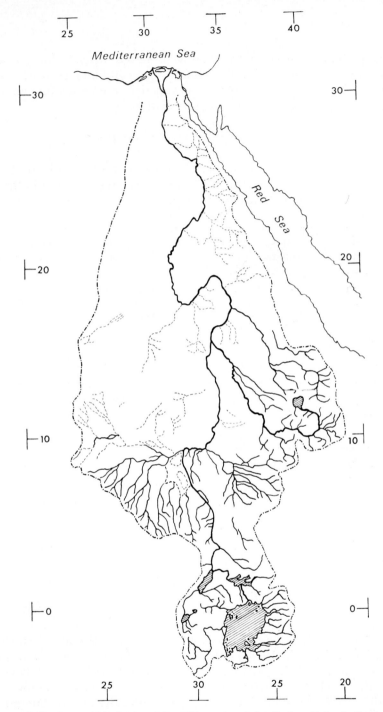

Fig. 14. The hydrographic basin of the Nile system; the limits of the drainage basin are indicated by a dashed line. Note the origin of the water supply almost entirely outside the two countries relying on the Nile water. Ethiopia and the Lake Victoria basin are the sources; the numerous rivers from the Congo (Zaire) divide are mostly lost in the swamps and their water supply is negligible; Adapted from Lockermann 1953 (unpublished thesis) and other sources.

of drought. This brings us to the total water resources available. Although the drainage area almost 3 million Km², the second largest in Africa and the fifth globally, yet 44% of this great area does not contribute anything to the water amount, because of desert conditions.

Water budget of the Nile System

Water cannot be created, it can only be conserved. Over at least 50 years the amounts of the water supply have been measured and within limits of fluctuations the mean values are known.

White Nile at stations	Mean Annual discharge in 10⁹ m³	Blue Nile at stations	Mean Annual discharge in 10⁹ m³
Lake Victoria	21	Lake Tana	4
Lake Kioga	22	Roseires	50
Lake Albert exit	23	Dinder	3
at Lake No	10	Rahad	1
Sobat mouth	13		
Malakal	23		
Gebel Aulia	23		

23 → At Khartoum 51 includ. losses

Joint Nile

Sabaloka (VI cataract)	74
Atbara	12
Semna Gorge	86

These data (Ministry of Irrigation Khartoum 1957) are the total natural supply of the river system including most of the losses. Note especially the difference of the discharge from the Albert Nile and that from the exit of the swamps at Lake No. The last figure at Semna represents the volume of water flowing now into the Aswan High Dam basin and Egypt. All figures are mean annual values from over 50 years of recording, but mean figures do not represent actual reality. Enormous fluctuations exist, – in 1913/14 the annual discharge at Semna was only 42 10⁹ m³; in 1878/9 it was 151. The seasonal differences are very large – during the flood 700 × 10⁶ m³ entered Egypt per day, during low water only 40 × 10⁶ m³. The dams alter partly the discharges but not the annual volume of water available. All the dams in the Sudan are seasonal storage reservoirs, but the new Aswan basin is a century storage reservoir to alleviate the fluctuations between the years. The operational scheme

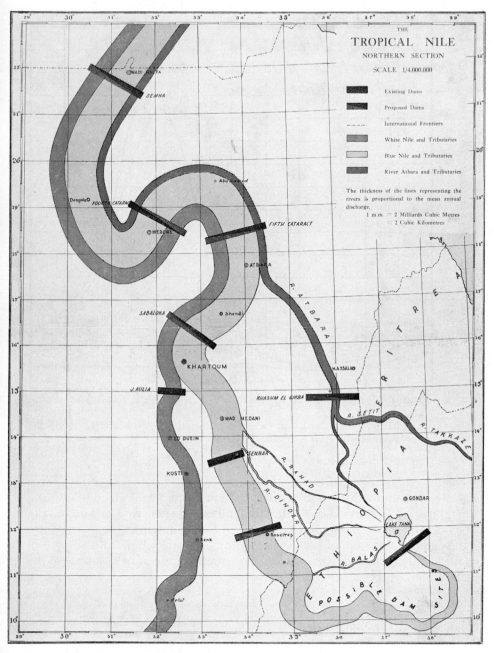

Fig. 15a. Graphic representation of the water supply by the main tributaries; the thickness of the belts represents the mean annual discharge, 1 mm = 2 × 10⁹ m³. Of the dams indicated in this proposal (1957) those at Tana, Sabaloka IV and V cataracts and at Semna have not been built and are at present obsolete. From Min. of Irrigation Khartoum 1957.

149

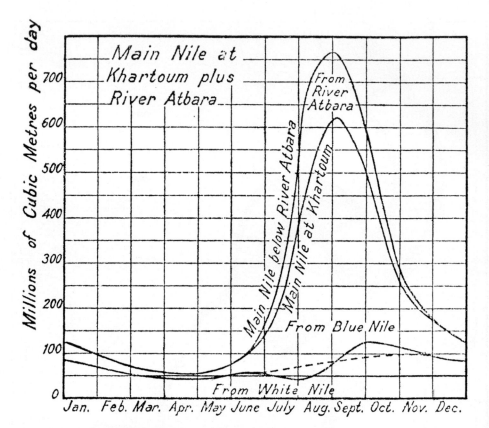

Fig. 15b. Discharges of the Nile and its main tributaries, averages 1912–1936; the long-term amounts of water are unchanged. (Hurst 1952).

of the dams in the Sudan are correlated so as to ensure a steady flow; the Sennar and Roseires dams cater for and are owned by the Sudan, the Gebel Aulia dam is Egyptian, though in Sudan territory.

A schematic map illustrates the annual contributions of the system graphically (fig. 15). Expressed as percentages of the annual supply of the particular rivers is:

At high level	Blue Nile	68%	{ (of which 7% from Lake Tana,
	White Nile	10%	{ 7%–10% Rahad and Dinder)
	Atbara	22%	
At low level	Blue Nile	17%	
	White Nile	83%	
	Atbara	0%	

150

Evaporation

Water loss by evaporation is a serious matter for an arid area. There is a controversy between the specialists as to the influence of vegetation on evaporation rates, and especially those from swamp-vegetation. Most data come from tank experiments enclosing a known space of open water and for comparison a similar area of swamp vegetation. Vegetation transpires and this varies with climate, especially temperature and relative humidity. Penman (1963) maintained on the basis of available references and his own brief experiments that 'the transpiration from the papyrus and evaporation from the open lagoon will be nearly equal'. Migahid (1952) who conducted such tank experiments in the 'Sudd' region in 1947, 1948 had different results depending on the climate of each month; on the whole there was almost no difference between the open water and tanks and those with papyrus. Hurst (in litt.) recalled experiments at Shambe, in the middle of the swamps, in large (10 m²) tanks which gave an evaporation rate of 1.25 in papyrus tanks in relation to the Piche Evaporimeter outside the tank. Further experiments in 1948/49 at the same place gave about 10% more water loss from vegetation than from open water; see also Vol. X of the Nile Basin. Very exact observations were made by Rijks (1969) in Uganda papyrus stands and he concludes: 'From the limited amount of material, it is concluded that evaporation from an old stand of papyrus, in which a fair proportion of brown and dried heads is visible, is about $60 \pm 15\%$ of the evaporation from open water'. His data show that only 70% of the net radiation was used in evaporation. The data by Rijks refer to undisturbed vegetation; in this connection it may be mentioned that Migahid (1952) noted that the state of the papyrus in his $6\frac{1}{2}$ years old tank was 'very poor as compared with the dense and luxuriant papyrus outside'. It seems therefore that the enormous water loss in the Upper Nile swamps, amounting to half of the discharge flowing into them, is not caused by the permanent papyrus stands but probably by spillage and evaporation of the temporary grasslands (see chapter on Upper Nile swamps).

Evaporation from open water is a very important hydrological feature; the following records (compiled from Hurst 1957) mean annual losses in metres of open water measured in tanks:

Lakes Victoria (Albert, Edward) 1.38 m	Lake Tana	1.09 (0.54–1.29)
Malakal	1.24	Roseires-W. Medani 2.30

Khartoum	2.85
Atbara-Wadi Halfa	2.77
Upper Egypt	1.64
Cairo	1.02
Delta	0.84

Oases	2.37
Faiyum	0.84

The influence of aridity is clearly seen with the highest evaporation recorded in desert climates. The losses by evaporation in dams are especially high and those of the Aswan High Dam basin will be enormous. A calculation by Peixoto *et al.* (1973) based on the full reservoir, covering 7,800 km², and assuming evaporation at 3 m per year, gives a loss of $24.10^9 m^3$, which would be a quarter of the total discharge.

These lossess will remain although large amounts of water may be 'saved' in upstream regions by the Jonglei and other schemes. Man has changed the ancient river profoundly and the extraordinary prophecy by a politician should be quoted. When the old Aswan Dam was completed in 1902 Winston Churchill wrote with great enthusiasm '. . . those giant enterprises may in turn prove but the preliminaries of even mightier schemes, until at last every drop of water which drains into the whole valley of the Nile . . . shall be equally and amicably divided among the river people, and the Nile itself . . . shall perish gloriously and never reach the sea' (quoted after Issawi 1963). This has finally happened.

The fact of this statement remains, some scepticism may be expressed as to its optimism. The ecological consequences of these and further man made changes are now discussed widely, e.g. by Kassas (1971).

REFERENCES

Chélu, A. 1891. Le Nil, le Soudan, l'Égypte. Librairie Chaix et Gorman, Paris.
Hamdan, G. 1961. Evolution of irrigation agriculture in Egypt. In: A history of land use in arid regions. ed. L. Stamp. UNESCO, Paris.
Hellström, B. 1951. The oldest dam in the world. Inst. Hydraulics, Roy. Inst. Tech., Bull. no. 20. Stockholm.

Hurst, H. E. 1952. The Nile, a general account of the river and the utilization of its waters. sec. edition (1957). Constable, London.

Hurst, H. E. & Phillips, P. 1932. Measured discharges of the Nile and its tributaries. The Nile Basin, vol. II; Physical Dept. Paper no. 28. Min. of Public Works – Govt. Press 1932; see also vol. VIII 1948.

Issawi, C. 1963. Egypt in revolution. Oxford Univ. Press, London.

Kassas, M. & Imam, M. 1953. On the vegetation and land reclamation in desert wadis. UNESCO Arid zone programme, Report NS/AZ/142.

Kassas, M. 1971. The river Nile ecological system: A study towards an international programme. Biol. Conserv. 4: 19–25.

Migahid, A. M. 1952. Further observations on the flow and loss of water in the 'Sudd' swamps of the upper Nile. Fouad I University Press, Cairo.

Min. of Irrigation, Khartoum 1955. The Nile Water Question. Survey Dept., Khartoum.

Min. of Irrigation, Khartoum 1957. Sudan Irrigation. Sudan Survey Dept.

Murray, G. W. 1947. A note on the el-Kafara; The ancient dam in the Wadi Garawi. Bull. Inst. d'Egypte 28: 33–43.

Peixoto, J. P. & Kettani, M. A. 1973. The control of the water cycle. Scient. American, April 1973: 46–61.

Penman, H. L. 1963. Vegetation and Hydrology. Tech. Comm. 53, CAB, Farnham Roy., England. Commonwealth Bureau of Soils.

Popper, W. 1951. The Cairo Nilometer. Public. Semitic Philol. vol. 12.; Univ. of California Press.

Rijks, D. A. 1969. Evaporation from a papyrus swamp. Quart. J. Roy. Meteorol. Soc. 95: 643–649.

Teclaff, L. A. & E. 1973. A history of water development and environmental quality. In: Goldman, G. R. et al. Environmental quality and water development. Free-mantle, San Francisco.

On a grand scale fig. 17a, b is the Delta and Faiyum built up from the dark Nile sediments over many thousands of years and intensely cultivated. Nile sediments have been used for tracing the phases of geological history of the river by mechanical, mineralogical and geomorphological appearance. It must be said at once that the present alluvial sediments overlie enormous deposits of former periods at least from the Miocene onwards (see ch. 1 and 2).

Already a number of ancient observers commented on the alluvial soil of Egypt and called it a 'gift of the Nile'. But exact studies started much later. Engineers of Napoleon's expedition estimated the deposition rate at 0.10 m per 100 years. More detailed investigations were carried out later

Fig. 17a. The Nile Delta from space; a wide panorama embracing Iraq, Saudi Arabia, Sinai, the Red Sea, the Suez Canal. The eastern and western deserts enclose the alluvial Delta, the richest agricultural area of Egypt and the most densely populated one. The Rosetta branch of the Nile is just visible; also visible are the four coastal lakes; courtesy of NASA; Gemini 4 S65–34776.

156

Fig. 17b. The Fayium Oasis and the Western part of the Delta with Wadi Natrun (black dots left centre); Note the intense cultivation and the black area of Lake Qarun; courtesy of NASA.

in the 19th and intensely in the 20th century. Most prominent are those of Ball (1939) and Simeika (1940), carried out in the Nubian and Egyptian part of the Nile. At present the total supply of sediments comes from the Blue Nile and the Atbara river with their head waters in Ethiopia. Hurst (1952) estimated the average annual load of sediments at Wadi Halfa at 100. 10^6 tons, over 90% coming down during the 4 months of the flood and only two million tons during the rest of the year. The sediment load rises from 100 g/ton of water at low river to between 2.5–4 kg per ton during the flood. Of the sediment arriving at the door step of Egypt (Ball 1939), 33% fell out between Wadi Halfa and Cairo, 14.5% were distributed by irrigation in Upper Egypt, 52.5% remained in suspension at Cairo and were transported to the Delta from August to November. The composition of the sediment at Wadi Halfa was averagely: sand 30%,

157

silt 40%, clay 30%. These conditions are now completely changed through the influence of the Aswan High Dam and also the Roseires dam on the Blue Nile. For the new sediment regime in the Wadi Halfa-Aswan region see ch. 19.

In the Sudan possible changes caused by the new (1966) Roseires dam have been studied and some pertinent results are quoted after Omer El Badri Ali (1972), with the author's permission. Conditions existing in 1967/69 after the dam closure were compared with those from 1959–61. Total figures for suspended matter passing Khartoum werein millions of tons: 1967 – 65, 1968 – 36, 1969 – 69.5; these values are lower than in the years before the operation of the Roseires dam. Hammerton has commented in ch. 16 on the deposition of some sediment quantities above the Roseires dam and their biological effects. It is obvious from the figures quoted that the sediment load varies from year to year with the discharge of the Ethiopian Blue Nile and this depends on the rainfall in the Highlands (see ch. 8). A graph (fig. 18) shows the annual fluctuation of the sediment load at Khartoum and its rise during the flood. Concentrations of the sediment load keep within the ranges observed by Simeika 50 years ago but their deposition along the Blue Nile may show changes. Of the total amount of sediment carried past Khartoum in the joint Nile 87% come

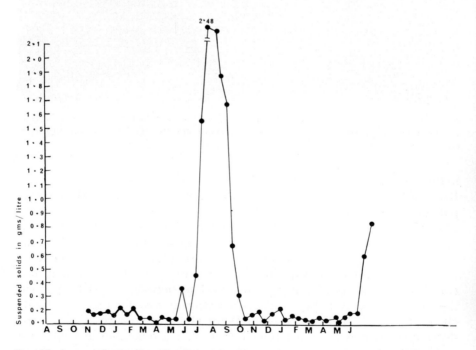

Fig. 18. Annual fluctuation of sediment load in the Blue Nile at Khartoum 1968–1970; from Omer Ali Bedri (unpublished).

from the Blue Nile, most of the rest comes from the Atbara; the White Nile contributes negligible amounts compared with the above.

Besides the suspended matter is the dissolved part; this will be examined in the chemistry of the Nile system (ch. 26). There remains the 'bedload' carried by the river along the bottom. Together with the finer sediments this contributes in the meanders of the Blue Nile to the formation of sand banks and islands. Air observation shows the continuous shift of these formations; these are on small scale repetitions of the great shifts and inundations which formed the Gezira plain in the past (see ch. 2 and 4a).

The mineralogical composition of Nile sediments was used by Shukri (1950) to trace the origin of palaeolithic sediment terraces. According to him the heavy mineral fraction of volcanic origin characterises the Blue Nile and even more the Atbara sediments. Magnetite, augite and epidote are present, the light fraction is composed of clay minerals, quartz and feldspar. These early investigations have been recently repeated and found on the whole correct. But the White Nile load, though very small and previously regarded as mainly of metamorphic origin (sillimanite and staurolite) has also a small admixture of volcanic provenience which is derived from the Sobat and its Ethiopian headwaters.

If at present Nile sediments come from the above mentioned sources it was not so in the past; with a more favourable rain regime sediments from previously active water courses in the north supplied sands and gravel of local origin as shown in ch. 1 and 2. Their presence together with wind-blown sands makes the unravelling of ancient sediments difficult; though the alluvial cover of the present regime is clearly recognisable.

These alluvial material accounts for the fertility of the cultivated land in the Nile valley. When deposited and dry the Blue Nile sediments are dark-grey, and form layers of friable, fine composition; they crack easily and are porous (fig. 19). Their nutritional value, once praised then disputed, has been found rich in phosphates and nitrates in laboratory tests and also in the field (Elster & Gorgy 1959). The rich bacterial flora of the river water is adsorbed to the sediment particles and sinks with them to the bottom (Jannasch 1956).

The fall-out of these sediments is of fundamental importance; it creates the soil for agriculture. Attia (1954) has examined the distribution and thickness of the alluvium in Egypt. Wherever the geological conditions allow for the Nile valley to broaden the Nile deposits its sediments; the width of this alluvial valley varies with the regions but on average extends from 2.8 km at Aswan to 15–17 km at El Minia and Beni Suef. Maximal extensions are up to 23 km. This is the classical land of the Nile valley. In the Faiyum oasis, once a lake of Nile origin (see ch. 4b), a clear lake delta spreading fan-wise can be recognised, showing the point of entry.

Fig. 19. Deposits of sediments on the river bank of the Blue Nile at Khartoum; Note
the layering and visible friability of the fine silt; courtesy of Omer Ali Bedri.

The average thickness of alluvial deposits is about 9 m in the Nile valley
but about 12 m in the Delta.

Recent calculations of rates of deposition of sediments in the Delta have
been slightly less than those of Napoleon's engineers and stand now at
0.09 m per century, but this is hypothetical in view of the fluctuations in
the sediment load over years. Worrall (1958) estimated the deposition
of silt in irrigated fields at Khartoum as 0.02 m per century. For the
latest data on Blue Nile sediments and their deposition see the Epilogue.

Fairbridge has reviewed Nile sedimentation in the last 20,000 years; in
the cataract section sediments were deposited extensively during 'dry'
seasons. The next flood flushed them out and ultimately most fine elements
went downstream.

This is now changed completely. The long basin created by the Aswan
High dam acts in the same way as all obstacles to free flow. The current
slackens and with it the carrying capacity of the water; now sediments
fall out at the upper end of the reservoir in the Wadi Halfa region. This is
treated in detail in ch. 19. The limnologist knows this phenomenon, it has
occured in every dam basin on the Nile. Sediments fall out, the water
clears, photosynthesis can act on rudiments of plankton with ultimate
great algal development and all other biological consequences. This is
already happening in the new dam few years after its construction.

REFERENCES

Attia, M. J. 1954. Deposits in the Nile valley and the Delta. Geol. Survey of Egypt. Govt. Press, Cairo. 356 pp.

Ball, J. 1939. Contribution to the geography of Egypt. Govt. Press, Cairo.

Elster, H. J. & Gorgy, S. 1959. Der Nilschlamm. als Nährstoffregulator im Nildelta. Naturwissenschaften 46: 147.

Fairbridge, Rh. W. 1963. Nile sedimentation above Wadi Halfa during the last 20,000 years. Kush 11: 96–107.

Hurst, H. E. 1952. The Nile, a general account of the river and the utilization of its waters. Constable, London.

Jannasch, H. W. 1956. Vergleichende bakteriologische Untersuchung der Adsorptions-wirkung des Nil-Treibschlammes. Ber. Limnol. Fluss station Freudenthal 7: 21–27.

Omer El Badri Ali. 1972. Sediment transport and deposition in the Blue Nile at Khartoum, flood seasons 1967, 1968 and 1970. M. Sc. Thesis, Univ. of Khartoum (unpublished).

Simaika, S. M. 1940. The suspended matter in the Nile. Phys. Dept. p. no. 40, Min. Publ. Works, Egypt.

Shukri, N. M. 1950. Mineralogy of some Nile sediments. Quarterly J. Geol. Soc. London 105: 511–534.

Worral, G. A. 1958. Deposition of silt by the irrigation waters of the Nile at Khartoum. Geogr. J. London 124: 129–222.

IV. HYDROBIOLOGY AND LIMNOLOGY OF THE MAIN PARTS OF THE NILE SYSTEM

$NO_3.N/l$. With other evidence for low concentrations of an alternative nitrogen source, $NH_4.N$, it seems possible that the nitrogen supply may be important in the regulation of the lake's fertility. The measured concentrations of soluble (reactive) phosphate-phosphorus were not particularly low, being above 7 $\mu g/l$ in the upper offshore layers during most of the year. The total phosphorus content was considerably greater, usually in the range 30–50 $\mu g/l$ in the upper layers. As a result of mineralization, the difference was usually less marked in the lower layers.

The concentrations of two other elements, iron and manganese, showed responses to stratification in line with their well known behaviour in lakes where the hypolimnia are seasonally depleted of oxygen. Accumulation took place in the oxygen-poor lower layers during strongest stratification, presumably as a result of chemical reduction there. The increased concentrations were, however, not very large (under 1 mg/l), in correlation with the limited distribution of very low concentrations of oxygen. When complete vertical mixing and well-oxygenated conditions were re-established by seasonal cooling, the deep accumulations of iron and manganese largely disappeared.

Nutrient chemistry and its relation to primary (phytoplankton) production in Lake Victoria has been examined in a series of papers by G. R. Fish, J. H. Evans and J. F. Talling and others in the last 25 years (see chapter 27). The primary production of the fringe of macrophytes is treated in chapter 11 by K. Thompson. Some investigations have been carried out on the chemistry of bottom sediments (ch. 25) and on their palaeolimnological features (ch. 3). The zoological record is uneven; zooplankton is known as to its species composition, to which early observers have contributed from 1888 onwards, when the first plankton net was dipped into Lake Victoria by Stuhlmann and Emin Pasha. Various expedition-type investigations completed the faunistic list. But only two papers (Worthington 1931 and Rzóska 1957, see ch. 24, have examined the distribution; we know nothing about densities and production problems. The benthos of the whole lake has, similarly, not been examined in a systematic way. But a rich and highly informative list of papers exist on mass emergences of mainly littoral insects, their numbers, generation cycles, importance as fish food (MacDonald, Tjönneland and Corbet) (see ch. 25). Excellent faunal lists exist on some groups especially the molluscs (Mandahl-Barth). Beadle has examined the physiological adaptations of some invertebrates to swamp conditions, Greenwood and others for some fishes. The fish fauna, famous for its 'adaptive radiation' is mentioned in ch. 7 by Greenwood. Some zoogeographical remarks on vertebrates are made in ch. 6.

More than eight million people live in the catchment area of Lake Victoria with an average density of 50/km², higher than in most of tropical Africa.

REFERENCES

Beadle, L. C. 1974[1]. The inland waters of tropical Africa. Longman, London.

Bishop, W. W. 1969. Pleistocene Stratigraphy in Uganda. Uganda Survey 10: 128 pp. Govt. Printer, Entebbe.

Fish, G. R. 1957. A seiche movement and its effect on the hydrology of Lake Victoria. Col. Off. Fisher. Public. 10, 68 pp., H.M.S.O.

Hurst, H. E. & Phillips, P. 1931. The Nile basin, vol. I, General description of the Basin. Meteorology, Topography of the White Nile basin. Govt. Press, Cairo.

Hurst, H. E. 1957. The Nile. Second rev. ed. Constable, London.

Kitaka, G. E. B. 1971. An instance of cyclonic upwelling in the southern offshore waters of Lake Victoria. Afr. J. Trop. Hydrobiol. & Fisheries 1: 85–92.

Krishnamurthy, K. V. & Ibrahim, A. M. 1973. Hydrometeorological studies of Lake Victoria, Kioga and Albert. UNDP/WMO Hydrometeor. Survey, Entebbe, Uganda.

Newell, B. S. 1960. The hydrology of Lake Victoria. Hydrobiologia 15: 363–383.

Rzóska, J. 1967. Notes on the crustacean plankton of Lake Victoria. Proc. Linn. Soc. London, 168: 116–215.

Talling, J. F. 1957. Diurnal changes of stratification and photosynthesis in some tropical African waters. Proc. Roy. Soc. B, 147: 57–83.

Talling, J. F. 1966. The annual cycle of stratification and phytoplankton growth in Lake Victoria (East Africa). Int. Rev. Hydrob. 51: 545–621.

Talling, J. F. & Talling, I. B. 1965. The chemical composition of African lake waters. Int. Rev. Hydrob. 50: 421–463.

Worthington, E. B. 1930. Observations on the temperature, hydrogen-ion concentration and other physical conditions of the Victoria and Albert Nyanzas. Inter. Rev. Hydrob. and Hydrogr. 24: 328–387.

Worthington, E. B. 1931. Vertical movements of freshwater macroplankton. Inter. Rev. Hydrob. and Hydrogr. 25: 394–436.

[1] This comprehensive work appeared to late to be considered in detail.

11. SWAMP DEVELOPMENT IN THE HEAD WATERS OF THE WHITE NILE

by

K. THOMPSON

Two broad kinds of swamps can be distinguished in the head waters of the White Nile of the East African Lake Plateau: seasonal (temporary) swamps and permanent. The general landscape and flora of this area have been described in Langdale-Brown *et al.* (1964) and Lind & Morrison (1974).

Seasonal swamps

Both seasonal and permanent types result from impeded drainage, but the former are restricted to land which is only periodically inundated or is subject to 'profile' waterlogging. Some 5,000 km² of grasslands and tree savannas of Uganda alone can be classified as seasonal swamp of one type or another. Most of this is associated with the heavy, impermeable, clay soils of the shallow Lake Kioga basin, where permanent swamp gives way to seasonal types and then to savanna woodland on higher ground.

As in the vast swamps of the Southern Sudan, many of the seasonal swamp grasslands provide valuable grazing for game and livestock during the dry season. All the grasses are perennial and, since shoots can be up to 3 m in height at the beginning of the dry season, it is common practice to maintain the quality of the fodder by burning 'over-mature' grasslands towards the close of each dry season. The 'wettest' of the seasonal swamp grasslands are dominated by varying proportions of *Echinochloa pyramidalis, Leersia hexandra, Oryza barthii* and *Cyperus latifolius,* and similar communities are associated with seasonal flood-plains throughout Africa. Slightly drier floodplain and valley swamps are dominated by grasses such as *Sorghastrum rigidifolium, Cynodon dactylon, Sporobolus pyrimadalis, Loudetia kagerensis, Setaria sphacelata, Brachiaria soluta* and *Imperata cylindrica.* A consequence of over-heavy grazing pressures and too-frequent burning regimes is that fast-growing, unpalatable, fire-tolerant grasses such as *Imperata* tend to increase their cover at the expense of the more desirable fodder grasses.

Except in regions with regular rainfall, tree cover is sparse on land with impeded drainage. Thus, in the Lake Kioga region, severe waterlogging rules out the growth of savanna trees and the length of the dry season is too great for forest development. However, the less-severely profile-waterlogged soils are colonised by *Acacia drepanolobium* (the 'whistling thorn'), the red-barked *A. seyal, A. sieberiana,* the broad-leaved *Combretum ghasalense* and the palm *Balanites aegyptiaca.* The northern and north-western fringes of Lake Victoria experience a much higher rainfall than the Nile basin to the north, and seasonal swamp forest becomes fairly common at slightly higher elevations than the permanent swamps. In fact, many permanent swamps are presumably seral to swamp forests comprising spp such as *Mitragyna stipulosa* (a tree with well-developed 'breathing roots'), *Macaranga schweinfurthii*, and the palms *Phoenix reclinata* ('wild date'), *Raphia monbuttorum* and *Calamus deeratus* (the climbing 'rattan cane'). At higher altitude (2,000 m +) in the Kigezi region of Uganda are seasonal swamp forests of *Syzygium cordatum*, which bear a remarkable structural similarity to the fen-carrs of higher latitudes. The conifers *Podocarpus milanjianus* and *P. usambarensis* are also high altitude species, but they are not found in swamps outside their single medium-altitude locality. Here, at the mouth of the Kagera River, the occurrence of both trees with *Baikaea insignis* makes their periodically-inundated forest type an edaphic curiosity.

Permanent swamps

Permanent swamps will only develop where shallow, moving water is perennial and protected from wave action and large changes in seasonal level.

The northern and southern shores are low-lying, with shallow profiles, and are deeply dissected into bays and inlets (see geological history in ch. 3). Run-off from land, stream and river inflow and lake turbulence maintain water movement (fig. 24).

In short, the northern and southern shores of Lake Victoria are ideally suited to supporting swamp and particularly so to papyrus swamp development. Large expanses of permanent swamp (mainly papyrus) are also found in the extremely shallow (average depth only 4 metres) and highly dissected Lake Kioga (level amplitude 0.35 m yr), where perennial through-flow is guaranteed by the Victoria Nile and the Mt Elgon torrents. In the western rift valley, Lake George has an exposed shore-line, but it is small in area and shallow. Its level is stabilised by the Kazinga Channel outflow to Lake Edward, and swamps are particularly well developed where the inflows from the Rwenzori Mtns deposit their sediment loads. Lake Edward, on the other hand, is steep sided and deep and, although the amplitude is only 0.5 m yr, wave-action and lack of a suitable substrate discourage swamp formation. The single outflow from Lake Edward, the Semliki River, enters Lake Albert, which possesses almost identical structural and hydrological characteristics. However, the delta formed where the Victoria Nile enters the lake and releases its silt load supports a large swamp. Considerable swamp development is also found along the slow-flowing Albert Nile.

Most of these lake and river swamps are dominated by papyrus.

178

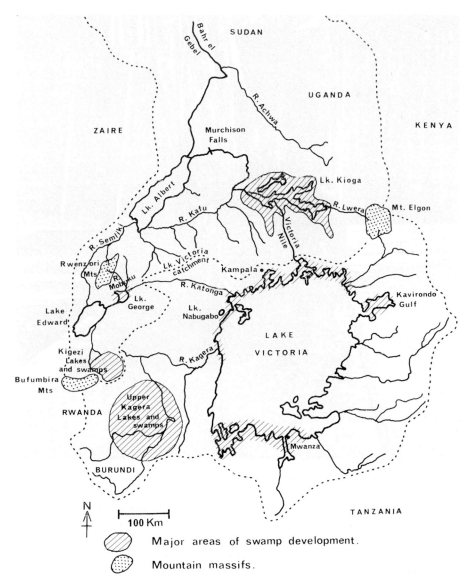

Fig. 24. The catchment area of the headwaters of the White Nile and (shaded) areas of swamp development around the shores of lakes in the East African plateau. K. Thompson.

Papyrus swamps

Cyperus papyrus L. is almost certainly native to the Palestine region, and it still grows in the Huleh swamps of the Jordan valley – the northern extension of the great rift valley system which flanks Uganda. Apart from

179

thetic bracteoles. Typically over 95% of the biomass of a papyrus swamp is attributable to the dominant plant itself. The interwoven roots and older, buried portions of rhizomes (fig. 25) form a compact floating mat up to 2 metres thick which, together with its living vegetation, typically weighs over 100 kg (wet weight) per square metre. The mat is not, however, very flexible, and severe wave-action or level fluctuations in excess of 1.5 m yr result in the separation of 'floating islands' of papyrus swamp. Particularly around the turn of this century, such islands were a serious navigation hazard in the Upper Nile Swamps of the Sudan. Newhouse (1928) observes that this period appears to coincide with a series of peak-discharge years for the Nile river. A similar hazard and correlation is cited by Camus (1957) for the Lualaba in Zaire. Even today drifting islands are troublesome in the Sudd region and on Lake Kioga. Most years the tugs of the East African Railways and Harbours Corporation are required to tow away papyrus islands obstructing the approaches to the Kenyan port of Kisumu. Situated within the Kavirondo Gulf in the north-east corner of Lake Victoria, Kisumu is often blockaded by up to 400 floating islands, some exceeding 2000 tonnes in weight, which have separated from the fringing swamps during peak lake level. Since the lake rose (permanently?) by 2 metres in 1963, entire new swamps have formed in many places, and the remains of the old ones can still be identified below the floating portions of the mats of the present swamps.

Figure 26 incorporates various types of papyrus swamp, and therefore demands some explanation. In exposed situations, the water-lily zone is absent and, where water movement is particularly low, the grass *Miscanthidium* expands its cover at the expense of papyrus. As illustrated, the papyrus zone is divided into four sub-types:

a. The fringing zone is distinguished by its smaller, more numerous culms and its typically higher community biomass. In the river sections of the Albert and Victoria Niles, fringing swamps are usually limited to the first two zones and are often themselves fringed by a zone of floating *Vossia cuspidata* or *Echinochloa stagnina*. These grasses have much more flexible rhizomes than papyrus, and can thus better withstand high current velocities and wave-action. Migahid (1952) also refers to this interaction between flow-rate and swamp vegetation type in the Sudan swamps.

b. The papyrus/fern zone is the closest approach to a 'typical' papyrus swamp. A superficial tangle of dead culms – about three times as many as there are living ones – limits the understorey to a little more than the shade-tolerant *Dryopteris striata*, although the aerial environment is more successfully contested: the upper canopy is shared with scramblers and twiners such as *Melanthera scandens*, *Mikania cordata*, *Cissampelos mucronata* and *Ipomoea rubens*.

c. The *Limnophyton* zone is characterised by stable pools with dense growths of blue-green algae (mainly *Microcystis*), and is only present in the largest swamps.

182

d. The intermediate papyrus zone is always present in natural lake swamps, but may be suppressed by human activity.

The geographical distribution of *Miscanthidium violaceum* appears to be limited to the higher rainfall region of the Nile headwaters. It is not found in the Sudan swamps, or in the Zaire basin, and is replaced by *Miscanthidium junceum* in southern Africa. In its ideal habitat, the community is almost mono-specific, but it is frequently mixed with the cattail *Typha domingensis* and the enormous reed-like grass *Loudetia phragmitoides*. On the more mesophytic sites, stands are less crowded, and many other species enter below the canopy to construct a community reminiscent of Tansley's 'mixed fen'. *Miscanthidium* is intolerant of deep flooding, and swamps are often dry enough to support *Sphagnum* spp (*S. subsecundum* agg.) and other poor-fen bryophytes.

The spiny shrub *Hibiscus diversifolius* (Malvaceae), the legume *Aeschynomene elaphroxylon* and *Ficus verruculosa* are the only woody plants common to mature papyrus swamp. The first two are frequently associated with the outer fringing zone, whereas the large shrubby fig is commoner in the intermediate zone and in papyrus-dominated valley swamps. In parts of the Okavango swamps in Botswana, the growth of *Ficus verruculosa* is so vigorous that previously navigable channels are now being rapidly blocked.

Along the southern shore of Lake Victoria, the natural mesophytic vegetation is savanna woodland ('miombo') and savanna. The northern fringe and the Victoria Nile banks above Lake Kioga support forest but, because of the intense cultivation in this region, a more accurate regional description of the present vegetation would be 'Savanna-like, derived from forest'. The (potential) forest type in this region varies from 'medium-altitude evergreen forest' *(Piptadeniastrum/Albizzia/Celtis)* in the north-west, to 'medium-altitude semi-deciduous forest' *(Celtis/Chrysophyllum/Chlorophora)* as the precipitation falls below 1,500 mm yr to the east of Kampala. The permanent swamps of the northern shores of Lake Victoria may either abut directly on to mesophytic forest (or, more usually, its man-modified derivative), or merge into the swamp forest represented on Fig. 26. *Phoenix* dominated swamp forest is the natural successor to papyrus swamp when the shore profile is very low, and wild-date palms also commonly establish themselves in the drier papyrus valley swamps.

As explained in ch. 3, the late Tertiary and early Pleistocene drainage system, to the west of what is now the eastern rift valley, has been reversed by tectonic activity to the extent that the tributaries have become 'drowned' valleys. The original drainage system is now crossed by three watersheds (other than that of the Zaire/Nile), and long, papyrus-choked valleys now slope gradually in opposing directions from the Victoria/Kioga, Victoria/George and Albert/Kioga divides. The vegetation type of these swampy valleys is similar to the intermediate zone of Figure 26, and it is not easy to establish reliable floristic differences between those

183

Table 1. Influence of different vegetation types on water chemistry. Data abstracted from Viner (1974), Table 2. His analyses for papyrus swamps agree well with those cited by Thompson & Gaudet (paper in prep.) for a much wider range of sites.

	Conduc-tivity μmho/cm	Chemical composition of water expressed as ratio of mg $1^{-1} \times 10$: conductivity in micromhos. (Means of 11 sites)										
		HCO_3^-	$SO_4^=$	Cl^-	CA^{++}	Mg^{++}	K^+	SiO_2	Fe	Mn	$Po_4\text{-}P$	$NO_3\text{-}N$
Papyrus swamps, throughflow	114	0.9	2.8	0.4	0.4	0.3	0.3	1.6	0.02	0.03	0.002	0.001
R. Semliki tributaries from Rwenzoris	85	4.0	1.8	0.3	0.6	0.4	0.4	2.4	0.02	0.01	0.02	0.03
North-east Uganda savannas, runoff	82	5.9	0.6	0.2	1.0	0.3	0.6	1.1	0.007	0.001	0.04	0.03

dissimilarities occur in respect of bicarbonate, iron, manganese, phosphate and nitrate, and these are explained later. However, the chemistry of the run-off from the over-grazed and regularly burned savannas to the north of Lake Kioga has nothing at all in common with that of papyrus swamp throughflow. The high phosphate value for the former is particularly distinctive.

The above comparison is intended, firstly, to illustrate the mobility of calcium, potassium, phosphate and nitrate when land is only sparsely vegetated and, secondly, to demonstrate the importance of swamps in controlling, or 'normalising', water quality throughout the upper Nile catchment. Lake Kioga is, of course, completely encircled by papyrus swamp, so the high levels of nutrients draining from the savannas are mopped up before they even reach the lake. The following example is intended to show that this really does happen: Nakivubo swamp is one of the numerous papyrus-choked inlets along the north shore of Lake Victoria. The treated effluent from the often over-loaded Kampala main sewage works enters the swamp down a narrow channel, but its distinct chemical signature is completely erased during its passage below over a kilometre of swamp into the lake (UNDP/WHO, 1970).

It is apparent, then, that swamps do modify the chemistry of the water flowing through them. Carter (1955) also noted the very low dissolved nitrate and phosphate levels and the enhanced iron, manganese and sulphate concentrations in papyrus swamp through-flow. However, the extent to which chemical adjustments occur is dependent upon the size of the swamp, its vegetation type, and the chemistry, velocity and volume of the inflow. Naturally, the sampling and analytical methods employed can also greatly influence the results obtained.

The chemical interactions occurring within swamps are extremely complex, but some aspects must be understood before biological inter-relationships can be appreciated. The pH of papyrus swamps is commonly within the range 5.5–6.5 but, where the nutrient status of the system is low, it can be much less. For instance, the water supply to the swamps around Lake Nabugabo (see Fig. 24) contains less than 0.2 meq/l of dissolved solids (this is represented by a conductivity of less than 15 micromhos/cm). The cations are taken up by the plants and hydrogen ions occupy the exchange sites on the clay lattice of the substrate. Flow rates are low and inundation rare, so the deep-rooted *Miscanthidium* dominates the swamp.

The pH range within papyrus swamps accounts for the low bicarbonate values and also permits the very high free-CO_2 concentrations which can develop in an organic 'soup' under anaerobic conditions and at high temperatures. Free oxygen is usually completely absent at only a few centimetres depth below the swamp water surface. Nitrate then becomes the oxidising agent in decomposition processes, and the redox potential falls to about $+200$ mV. Although the ammonia formed by this redox

187

effective means is there of minimising production losses through parasites, disease and declining photosynthetic capacity with age than by keeping the photosynthetic organs young and entirely replacing them every three months?

Conservation

If a picture has now emerged of the Nile swamps as very productive ecosystems functioning at high water-, chemical- and energy-use efficiencies in a basically hostile environment, this picture was intended. Is there any man-directed ecosystem which can perform all the functions of a tropical papyrus swamp more efficiently, or even as efficiently? Conservation of swamps, as refuges of animal and plant life, has recently been brought into focus by various agencies, such as the International Union for the Conservation of Nature. It is, of course, most desirable that selected ecosystems should be totally conserved if possible, but in most cases we must accept that careful management of something similar to the original ecosystem is the most we can expect. Countries such as Egypt, Sudan, Ethiopea, Uganda, Rwanda and Burundi need water for domestic and industrial use, and land for agricultural development. The rivers provide water and the swamps have the greatest potential for agriculture and, regardless of the opinions of the conservation lobby, both resources will be greatly exploited during the next few decades. Funds and qualified personnel must be made available to ensure that swamp reclamation and water development projects in Africa in general, and the Nile basin in particular, take account of the distant future, and not just the next 10 or 20 years. Most development agencies will contend that this aim is already incorporated into current planning, but the facts do not often bear this out. Traditional agricultural practices in Africa acknowledge the scarcity of soil nutrients and the importance of the seasonal floods and the swamps in maintaining a stable system. Modern development technologies, such as flood control measures, high-yielding economic crops, expensive fertilizers, etc, have been imported from the Temperate Zone, and are too often applied in the tropical environment by agricultural engineers and hydrologists who have little appreciation of ecosystem dynamics and inadequate regard for long-term side-effects. It is important that they should see swamps not just as weed growths blocking communication channels and covering good agricultural land, but as a vital natural resource regulating water and nutrient flow, extracting silt and supporting fisheries. The swamps and bogs of Lesotho were the life-blood of the Orange River of South Africa, but man-induced damage to the catchment vegetation has been so great that flood-damping has been all but eliminated and erosion is extensive (Jacot-Guillarmod 1969). The swamps of the central African plateau are the key to the survival of the River Nile.

194

Acknowledgement

Drs. J. J. Gaudet (Kenyatta Univ. Call., Kenya) and A. B. Viner (Univ. of Malaya) have kindly provided unpublished material for this chapter. Dr. Viner has also supplied valuable comment on the draft.

REFERENCES

Banage, W. B. 1964. Some aspects of the ecology of soil Nematodes. E. Afr. Acad. 2: 67–74.

Banage, W. B. 1966. Survival of a swamp Nematode (Dorylaimus sp.) under anaerobic conditions. Oikos 17: 113–120.

Beadle, L. C. 1974. The Inland Waters of Tropical Africa: an Introduction to Tropical Limnology. Longman. (This summarises most previous work).

Bishai, H. M. 1962. The water characteristics of the Nile in the Sudan, with a note on the effect of *Eichhornia crassipes* on the hydrobiology of the Nile. Hydrobiologia 19: 357–382.

Bishop, W. W. 1958. Raised swamps of Lake Victoria. Geol. Survey Uganda, Records (1955–1956): 1–10.

Bowmaker, A. P. 1960. Seasonal hydrological changes affecting the ecology of Luaka Lagoon, Lake Bangueulu: some preliminary observations. CCTA/CSA Tech. & Sci. Publ. 63 (Proc. Lusaka Sympos.): pp. 22.

Calder, E. A. 1959. Nitrogen fixation in a Uganda swamp soil. Nature 184: 746.

Camus, C. 1957. Problèmes posés par les papyrus au Lualaba supérieur. Acad. Roy. Sci. Brux., B. Séances, 3: 1164–1185.

Carter, G. S. 1955. The papyrus swamps of Uganda. Heffer, Cambridge.

Deuse, P. 1966. Contribution à l'étude des tourbières du Rwanda et du Burundi. I.R.S.A.C., Rapp. 30 (1960–1964): 53–115.

Eggeling, W. J. 1935. The vegetation of Namanve Swamp, Uganda. J. Ecol., 23: 422–435.

Goma, L. K. H. 1960. The swamp-breeding mosquitoes of Uganda: records of larvae and their habitats. Bull. Ent. Res. 51: 77–94.

Goma, L. K. H. 1961. The influence of man's activities on swamp-breeding mosquitoes in Uganda. J. Ent. Soc. Afr. 24: 231–247.

Golterman, H. 1973. Natural phosphate sources in relation to phosphate budgets: a contribution to the study of eutrophication. Water Res. 7: 3–17.

Greenwood, P. H. 1955. Reproduction in the catfish *Clarias mossambicus* Peters, Nature 176: 516–518.

Greenwood, P. H. 1956. Reproduction of *Clarias mossambicus* in Lake Victoria. CCTA/CSA Publ. 25; II Symp. Afr. Hydrob. & Inland Fisheries.

Hurst, H. E. 1952. The Nile, (second edition). Constable, London.

Jacot Guillarmod, A. 1969. The effect of land usage on aquatic and semi-aquatic vegetation at high altitudes in Southern Africa. Hydrobiologia, 34: 3–13.

Jonglei Investigation Team. 1954. The Equatorial Nile Project and its effects in the Anglo-Egyptian Sudan, I–IV. Sudan Govmt, Khartoum.

195

water from Lake Albert, carried away in changing amounts by the Nile, influences its conductivity in a fluctuating way (Beauchamp 1956).

Lake Albert has an area of 5500 km², is 175 km long and 40 km wide; its maximum depth is about 50 m. The bathymetric map (Fig. 31) shows the steep slopes on the western side bordering the Ruwenzori. The lake differs in origin (ch. 3) and biologically from both Kioga and Victoria; the fish fauna is much poorer and without the rich speciation; Greenwood explains the historical reasons (ch. 7). The zooplankton has been studied by Green, the phytoplankton by Talling and these will be treated in the general chapters 24, 26 and 27. Worth mentioning is Green's discovery of a *Parabathynella* in its eastern shore psammon; see also Beauchamp (1956, 1963) and Worthington (1929).

The Victoria Nile enters the lake in its north-eastern corner and leaves

Fig. 32. Nimule, the end of the Albert Nile; Note the narrow gorge descending to the Sudan; air photo, Sudan Survey.

200

the lake a little further north, another feature influenced by its history. The Albert Nile flows north in a swampy valley with a very small slope of 0.022 m/km, until the escarpment is reached at Nimule and the river enters the plains of the Sudan (Fig. 32). Into this northern stretch enters the Aswa river which harbours and interesting fish fauna on which **P. H. Greenwood** has commented in ch. 7.

On descending the Uganda escarpment the Nile traverses a region of rocky incised valleys and more open country with rapids; the Arabic name of Bahr el Gebel, river of the rocks, fits this sector. Little is known at present of river life here; rocky, seasonal streams exert some influence on the chemical composition downstream (Talling 1957a).

The river now enters a large basin surrounded in a wide almost complete circle by higher ground above 500 m (Fig. 33), with hills and mountains. The slope of the Nile levels out (Fig. 4); Bor is at 419 m and

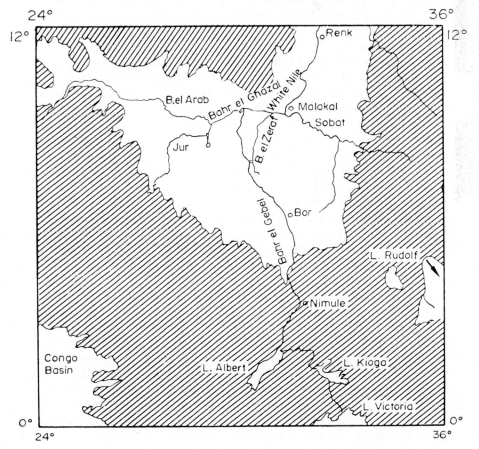

Fig. 33. The depression basin of the Upper Nile swamps in the Sudan; shaded area is over 500 m a.s.l., Scale 1 : 20,000,000; from Rzóska 1974.

Lake No at 386 m a.s.l., which gives an average gradient of 0.052 m/km over a stretch of 627 km of river. Currents slow down from an average of 1.13 m/sec at Bor to 0.61 m/sec at Lake No. After the rocky stretch the river becomes again bordered by swamp vegetation. Two main sectors can be distinguished northwards; from Juba to Mongalla and Bor, there is a relatively narrow fringe of grassland swamp with *Echinochloa stagnine* and *Phragmites* spp. dominant along the river, and *Echinochloa pyramidalis* in the flood plains. From Bor to Lake No this fringe broadens and is dominated by *Cyperus papyrus*. Here are the great swamps stretching from 6° to 9° lat. N., a very specific biome. Rainfall in this region is narrowed down to one season (see Fig. 10).

The Upper Nile Swamps

The Arabic name 'Sudd' used for this region means 'Blockage' and refers to the obstacles to navigation caused by the swamp vegetation. Some of the many descriptions of this region and the difficulties of travel have been mentioned in chapter 4d.

These swamps are predominantly riverain in sharp contrast to those around Lake Victoria treated in the preceding chapter. There is a spate of publications on the hydrology, soils, botany and general problems of this region starting in the 20th century with the report by Sir William Garstin (1901). Our present knowledge derives above all from the great work published by a team of the Sudan Government, the 'Equatorial Nile Project, vol. 1–V' (1954). This dealt with the possible effects of the proposal to bypass the main swamps by a canal from a place called Jonglei to the White Nile south of the Sobat mouth. This project which would avoid the enormous water losses in the swamp area (see ch. 8) is now near realisation. It will change the ecology of these great wetlands profoundly and affect the life of the Nilotic people. The report deals with almost every natural component of this biome.

Hydrobiological investigations by the University of Khartoum started here in a systematic way in 1953 after previous reconnaissance in 1947/8. A number of scientific contributions were published and some of the results are summarised by Rzóska (1974). Several factors combine to create this biome; a flat land surface with only a micro-relief, the impermeability of the widely distributed clay soils, and a peculiar hydrological regime. Rainfall (see Fig. 10) is strongly seasonal from June to mid-November, the river level rising, due to the rains in the Lake Victoria basin, with a peak in April–May, and to a low slope of the river bed. The Nile (Bahr el Gebel) and other swamp rivers meander and braid, winding through the swamps; some of the minor rivers especially in the Ghazal region lose their identity completely and never reach the main river.

The origin of the swamp region, still only hypothetical, is commented upon in ch. 2. Generally four elements form the landscape – the perma-

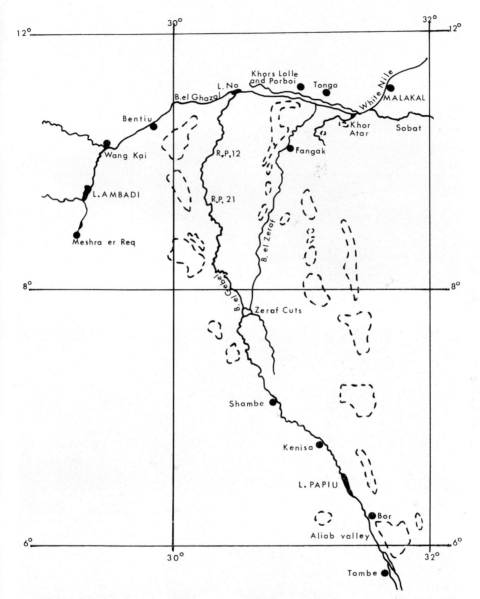

Fig. 34. Sketch map of Upper Nile swamps, encircled areas are the higher ridges with permanent villages; from Rzóska 1974.

nent swamps fringing the rivers, the flood plains with grasslands and the dry land ridges only few metres above the flood plains (Fig. 34); interspersed is a multitude of standing waters of varying size.

The permanent swamps show a very similar plant composition as those in Uganda. From the many descriptions of this flora the main components

203

Fig. 35. Papyrus wall on Lake No with a stand of *Vossia*; Photo Rzóska.

Fig. 36. Small inlet in swamp fringe with *Nymphaea* and *Trapa*; Photo J. Talling.

are compiled: *Cyperus papyrus, Vossia cuspidata, Phragmites mauritianus, P. Karka* and possibly *P. communis, Typha domingensis*; these form a fringing wall up to 5 m high (Fig. 35). A tangle of climbers is present, e.g. *Vigna nilotica, Luffa cylindrica, Cissus ibuensis* and *Ipomoea aquatica*. In inlets and bays (Fig. 36) floating and submerged plants flourish with *Nymphaea lotus*, *Najas pectinata*, *Potamogeton* spp., *Trapa natans*, *Pistia stratiotes*, *Ceratophyllum demersum*, the fern *Dryopteris gongyloides*, *Azolla nilotica*, *Utricularia* spp., *Ottelia* spp., and others. Of special interest is the recorded presence of *Eichhornia natans* which is an inconspicuous plant and quite different from *E. crassipes* which has disastrously invaded the Nile in 1957–1958 (see ch. 21). Previous to this invasion the papers by Chipp (1930), Andrews (1948), -igahid (1947), and McLeay (1953) should be consulted for details; see also ch. 5.

Trees are few and occur on the higher ridges and drier lands; the Ambatch tree, *Aeschynomene elaphroxylon*, is found in the swamps and is sometimes used for canoe rafts because of the lightness of its wood.

A detailed study of the relations between hydrological conditions, current and depth of flooding, and the distribution of vegetation was made by Sutcliffe (1973, 1974). Although it concerned the Aliab region at the beginning of the swamps, its results are valid also further north. The main components of the swamp flora are dependent on some defined conditions. *Cyperus papyrus* with its rhizomes and soft roots cannot penetrate hard soils and is confined to permanently moist positions along river and standing waters. More inland, soils dry up to a hard consistency during the rainless season of November–May. Also, papyrus does not tolerate large variations of water level; this incidentally excludes it from appearance in the Blue Nile. In the Bahr el Gebel changes in river levels are damped by the influence of the numerous lagoons and lakes adjacent to the river; they fall from 1.37 m at Mongalla to 0.78 m at Bor and 0.65 m at Shambe. Similarly currents slow down to a rate acceptable by papyrus. The distribution of papyrus reflects these two factors clearly. *Vossia cuspidata* is better adapted to stronger currents and can advance into the free channel of the river by submerged rhizomes, which are strong and flexible; pure stands of *Vossia* prevail in the upper parts of the swamps and intermingle with papyrus in the middle of the swamps. *Phragmites* spp. reach into the temporary inundated lands, being resistent to dry conditions, similarly *Typha* and *Echinochloa pyramidalis*.

The swamps in the Ghazal system differ – papyrus is stunted, low reeds replace the tall fringing walls of vegetation. Further differences are mentioned below.

The area of the permanent swamps along the Bahr el Gebel has been estimated from aerial photographs in 1930 at about 8000 to 10000 km^2; to this should be added large swamps in the other rivers. Each year flooding occurs due to the factors mentioned; the river spills through the alluvial banks into the slightly lower plains. The flood advances from the river inland to a distance up to 80 km, depending upon the relief (Fig. 38) and the resistance of the grass cover. These *temporary wetlands* are called 'toiches' by the Nilotics; they form the grazing grounds. The area of these flood plains has been estimated at about 70,000 km^2 with large

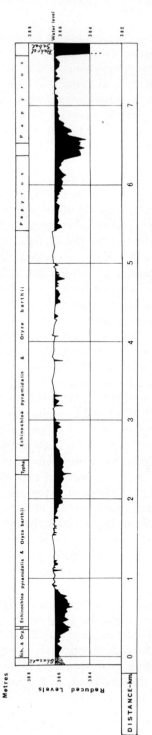

Fig. 37. Schematic cross section through Upper Nile swamps from Bahr el Ghazal to the Bahr el Gebel; Note micro relief and interspersed standing waters; from Jonglei Report (E.N.P.) A 21, 1954.

additions outside the central part. The third element, ridges of firmer land (see Fig. 34), are the sites of permanent villages of the Nilotics, Nuer and Dinka mainly. Further inland follows a dry 'Acacia-tall grass savanna'.

Man has adapted to the harsh conditions imposed by this region. The Nilotics are pastoral and have to rely on grazing, provided mainly by *Echinochloa pyramidalis* and *E. stagnina* and *Oryza barthii*; these can exist only where flooding is below 1 m. In 1954 about 700,000 people lived in the swamps with about 900,000 cattle and sheep. Grave political disturbances in recent years have upset conditions in the southern Sudan.

Life of the Nilotics is closely adapted to the annual cycle of the flood affecting their grazing grounds. Grasses in these flood plains grow tall and hard during the flood and are fired (Fig. 38); with the flood receding the tussocks sprout again and provide suitable fodder. The Nilotics follow the receding flood with their animals until they reach the permanent swamp fringe, where water is available throughout the rainless five or six months of November until May; temporary camps are established here (Fig. 39). The back migration to the permanent villages starts with the next flood in June–July. Each section of the population, usually a sub-tribe has to have a slice of the land with all three elements, dry land, 'toich' and access to permanent water; this old established set up is guarded zeal-

Fig. 38. Swamp firing; one of the pastoral practices used by the Nilotic cattle owning tribes to secure fresh growth from inundated grass lands behind the papyrus fringe; Photo P. Gay.

Fig. 39. Nilotic cattle camp on the river bank; this is part of the pasture migrations during the dry season when the inland country dries up; Photo P. Gay.

ously. The cultivation of crops is hazardous because of insect pests and ravages by mass appearances of *Quelea*, a weaverbird; biting insects, mosquitoes, tabanids and *Stomoxis* affect the life of man and beast severely (see ch. 23). All in all this is a difficult environment for man. For details see Rzóska (1974).

The non-human ecology of the swamps has attracted many observers in the past, especially in the 19th century (see ch. 4d.). More recently, practical considerations have come into the foreground, especially hydrological problems and agricultural improvement. Numerous studies have been devoted to this region culminating in the volumes of the 'Equatorial Nile Project'. In contrast, the hydrobiology of the swamps was largely unknown before the work of members of the University of Khartoum, established in 1946. But it was 'expedition' type of work as compared with the station-based studies on Lake Victoria; the Upper Nile swamps are 1000 km from Khartoum and required a long journey. There is also a difference in the object of studies; general hydrobiology was the main object of the Khartoum team.

The swamp rivers and their limnological characteristics have been studied in five longitudinal surveys from 1953 to 1964 (Talling 1957a, Bishai 1962, Kurdin 1968 and Monakov 1969). In the two most extensive of these surveys (in 1954, 1955) from Lake Victoria to Khartoum, over a distance of 2,600 km, Talling has demonstrated in a series of graphs the influence of the swamps on the White Nile. In the middle of the swamps dissolved oxygen drops to a low level and fish death was reported from this region near Lake Papiu; pH fell, carbon dioxide increased, similarly iron (Fe), and sulphates were largely removed. Most of these conditions became more 'normal' again in the northern, final part of the swamps. Later investigators confirmed Talling's results and the general trend of limnological development. However, fluctuations of a number of limnological characteristics, measured at a single point of a continuously moving water-mass in a river, have been recorded at time intervals – see Rzóska (1974). For example, near Lake No the White Nile showed variations in conductivity from 190–550 μmho/cm, pH from 7.1 to 7.8, transparency varied from 0.32 m to 1.30 m, carbon dioxide from 3.8 to 15 mg/l, oxygen saturation from 17 to 65%; similar fluctuations were recorded in the other rivers. It should be remembered that rivers collect influences from afar; the Nile is subject to 'flushes' from Lake Albert with a chemical composition different from Lake Victoria; the Sobat is influenced by the Ethiopian High Plateau and the Ghazal river by the ironstone country of its western basin.

An incident, throwing light on decomposition process in the swamp Nile, has been recorded in an unexpected source. In 1910 a steamer in the Bahr el Gebel (White Nile) struck a 'sudd' block which liberated a large amount of 'marsh gas' (CH_4) which, ignited by the wood burning engine fire, severely burned the fireman and blistered the steamer paint (Hill 1972).

Standing waters interspersed in the swamps are usually in contact with rivers. They show, on the whole, a greater stability of water characteristics than the rivers, as the water renewal is very much slower. All the examined standing waters are shallow, not exceeding 2–3 m. They are not turbid like rivers, oxygen conditions are on the whole more stable and more favourable with, however, variation both diurnal and spatial; nutrients in a standing water can be used in a localised way.

The biological effect is most clearly visible in the development of phyto- and zooplankton. The phytoplankton in standing waters is dominated by Cyanophyceae (bluegreen algae), especially by forms of *Anabaena flos-aquae* and *Lyngbya limnetica*, whereas in rivers diatoms, especially *Melosira granulata* prevail. The quantities recorded are much higher in standing waters, ranging at the time of investigations between 1,700–3,600 units (cells, filaments and coils) per ml; in the scanty suspension of the rivers only about 40–140 per ml were present. Water-blooms occurred in some standing waters; in a lagoon separated from the main river by a papyrus wall, Talling (1957b) counted 131000 filaments of a bluegreen algae per ml with a high primary production (see ch. 27).

Similarly, zooplankton increased in standing waters to up to 260

Crustacea per litre, while in rivers only 1–2 Crustacea and few Rotifera per litre were found in numerous sampler catches.

A further difference, generally known, is that changes of plankton composition and density in standing waters are mainly temporal, whereas in rivers a floating water-mass carries a zooplankton of spatially varying composition and density past the observers point. Examples of these are given by the writer (Rzóska 1974).

Rivers collect characteristic features of water composition on their way. The best example of river individuality is the Bahr el Ghazal. It issues as a recognisable entity, of 200 km run, from Lake Ambadi, a shallow basin in the low reed swamps, the confluence of many swamp rivers from an ironstone basin, largely lost as discrete units. Lake Ambadi is transparent to the bottom (2.5 m) and bears a luxuriant submerged plant cover of *Ceratophyllum*, *Myriophyllum*, *Vallisneria* and *Utricularia*; it has a low mineral content and (at the time of investigation) a low pH, a different shore flora with little and stunted papyrus (Fig. 40). It has a pond plankton of a composition unlike that of any other investigated swamp lake of this region and above all an extraordinary richness of Desmids (Grönblad *et al.* 1958). The desmid flora of about 205 species and forms, many of them new, contrasts with a further collection downstream the Ghazal river with only 70 species and forms (Grönblad 1962), and above all with the rest of the Nile system with only 24 common species of desmids

Fig. 40. Swamps on the Ghazal river; Note the absence of papyrus and floating plants; Photo J. F. Talling.

recorded (Brook 1954). This poor picture prevails once the Ghazal river reaches its end at Lake No.

The richest habitat in the swamps is the free water of the vegetation fringe. Inside the vegetation conditions are less favourable. Transects by Talling (in Rzóska 1974) at two places, 500 km apart, show from inside to outside a rise of pH, a fall in CO_2 and a rise of oxygen saturation from as low as 5–10% to about 30–60%, and a fall of iron; but alkalinity, nitrates, ammonia, phosphates show general increases from inside to the fringe and some components show higher values here than in the free river, except pH. A closer examination of oxygen conditions in the vegetation fringe has revealed a wide range of saturation, with supersaturation in sites exposed to the sun in submerged or floating vegetation.

The favourable conditions of this outer fringe comprise also a drastic reduction of current as compared with midstream (Migahid 1947, Rzóska 1974), with shelter provided by inlets and bays, a richer supply of nutrients and food from the influx of organic matter from the vegetation (see ch. 11).

Invertebrates abound: 27 genera of Cladocera are represented by 34 species, a number of them circumtropical, similar to the Uganda swamps (see Thomas 1961); numerous Cyclopids, Ostracods, Hydracarina can be found as well as insect larvae and adults. Oligochaeta, some tube building, and Bryozoa, *Urnatella* and *Cordylophora* have been found (Rzóska 1949, (1974). Key animals, because of size and numbers, are the herbivorous shrimps, *Caridina nilotica* and the rare *Palaemon niloticus* and *Cyclestheria hislopi*, a detritus feeding Conchostracan.

The fringe is populated by large shoals of fishes, both juvenile and adults; some species are characteristic for this fringe habitat. Sandon (1951) has recorded here 17 species of 11 families.

The fish fauna of the Nile system is treated in a special chapter by H. Greenwood; but fisheries problems should be mentioned. Fishing is done either by professional fishermen, northern Arabs or Shilluk; some sub-tribes, like the Monythany Dinka, live permanently on the swamp edge near Bor (Fig. 42). The bulk of the swamp Nilotics fish occasionally, either communally or individually. Nilotics know their natural environment, especially grasses; they also know some 40 species of fishes and have their own names for them (Girgis 1948). The fishing potential is considerable (E.N.P. I, p. 345 and III 385). Many of the Nile fishes migrate locally to the warm inundated flood plains for breeding, according to Stubbs (1949). The finer distribution and details of their migrations are little known.

Two specific studies on microhabitats have been made by I. B. W. Thornton, who was a member of the University of Khartoum. One (1957) examined the faunal succession of invertebrates in the umbels of papyrus from the young, green stage to the old and dry end of development. Sap sucking insects were gradually succeeded by groups with

211

Fig. 41. Fisherman's dwelling in the swamps north of Bor; a sub-tribe of the Dinka gain their livelyhood from fishing; Photo J. Rzóska.

Fig. 42. Elephants in the swamps north of Bor during the dry season; like man, many animals have to migrate from the dry interior to permanent water; Aero Films.

212

different food requirements until only rapacious species remained like spiders etc. The other study dealt with the fauna of *Pistia stratiotes*, with a rosette of leaves floating above the water surface and a large bunch of submerged roots. Up to 300 individuals of 16 different invertebrate groups were found; a summary of this unpublished work with some faunal lists, including new Coleoptera, can be found in Rzóska (1974).

Although the swamps are monotonous and seem to be desolate, yet its permanent waters attract a great number of mammals, when the surrounding savanna dries out (Fig. 42). Of the former abundance, however, referred to in chapters 4d and 6, only much fewer numbers remain, and even these are threatened. The two antelopes, the Nile Lechwe and the Sitatunga, with hooves spread in adaptation to the soft grounds, are specific for the swamps. Over 60 species of birds live here; crocodiles have become scarce: some details are treated in the relevant parts of ch. 6.

REFERENCES

Andrews, F. W. 1948. The Vegetation of the Sudan. In: Tothill, Agriculture in the Sudan. Oxford Univ. Press.

Beauchamp, R. S. A. 1956. The electrical conductivity of the headwaters of the White Nile. Nature 178: 616–619.

Beauchamp, R. S. A. 1963. The Rift valley lakes of Africa. Verhdlg. Intern. Ass. Limnology 15: 91–99.

Bishai, H. M. 1962. The water characteristics of the Nile in the Sudan etc. Hydrobiologia 19: 357–382.

Brook, A. J. 1954. A systematic account of the phytoplankton of the Blue and White Niles at Khartoum. Ann. Mag. Nat. Hist. 12: 648–656.

Chipp, T. F. 1930. Forests and plants of the Anglo-Egyptian Sudan. Geogr. J. 75: 123–143.

Equatorial Nile Project. 1954. (E.N.P.) Its effects in the Anglo-Egyptian Sudan. Being the Report of the Jonglei Investigation Team vol. I–V. Sudan Government.

Garstin, Sir William. 1901. Report on Irrigation Projects on the Upper Nile, etc., accomp. a despatch by Lord Cromer. Part II contains a section: The Sudd. Publ. as 'Blue Book'.

Girgis, S. 1948. A list of common fish from the Upper White Nile with their Shilluk, Dinka and Nuer names. Sudan Notes and Records 29: 2–7.

Grönblad, R., G. Prowse & Scott, A. M. 1958. Sudanese Desmids. Acta Bot. Fenn. 58: 1–82.

Grönblad, R. 1962. Sudanese Desmids II. Acta Bot. Fenn. 63: 3–19.

Hill, R. 1972. A register of named power driven. craft, commissioned in the Sudan, II. Sudan Notes & Records 53: 204–214.

Kurdin, V. P. 1968. (in Russian) Data of hydrological and hydrochemical observations on the White Nile. Biologia Vnutrennich Vod, Inform. Biuletin no. 2. Acad. Sc. USSR.

MacLeay, K. N. G. 1953. The ferns and the fern allies of the Sudan. Sudan Notes and Records 34: 286–298.

Migahid, A. M. 1947. An ecological study of the 'Sudd' swamps of the Upper Nile. Proc. Egypt Ac. Sc., vol. III: 57–86.

Migahid, A. M. 1948. Report on a botanical excursion to the Sudd region. I excursion 1946. Fouad I University Press, Cairo.

Monakov, A. V. 1969. The zooplankton and zoobenthos of the White Nile and adjoining waters of the Republ. of the Sudan. Hydrobiologia 33: 161–185.

Rzóska, J. 1949. Cordylophora from the Upper White Nile. Ann. Mag. Nat. Hidt. 12: 588–560.

Rzóska, J. 1974. The Upper Nile Swamps, a tropical wetland study. Freshwat. Biol. 4: 1–30.

Sandon, H. ed. 1951. Problems of fisheries in the area affected by the Equatorial Nile Project. Sudan Notes and Records 32: 5–36.

Sutcliffe, J. 1973. Ph.D. Thesis, Univ. of Cambridge; ch. 4 The ecology of the vegetation of the floodplains.

Sutcliffe, J. 1974. A hydrological study of the southern Sudd region of the Upper Nile. Hydrol. Sciences – Bull. des Sciences Hydrologiques 19: 237–255.

Stubbs, J. M. 1949. Freshwater Fisheries in the northern Bahr el Ghazal. Sudan Notes and Records 30: 245–251.

Talling, J. F. 1957a. The longitudinal succession of water characteristics in the White Nile. Hydrobiologia 11: 73–89.

Talling, J. F. 1957b. Diurnal changes of stratification and photosynthesis in some tropical African waters. Proc. Roy. Soc. B 147: 57–83.

Thomas, I. F. 1961. The Cladocera of the swamps of Uganda. Crustaceana 2: 108–125.

Thornton, I. W. B. 1957. Faunal succession in umbels of Cyperus papyrus on the Upper Nile. Proc. Roy. Entom. Soc. Lond. 32: 119–131.

Worthington, E. B. 1929. The life of Lake Albert and Lake Kioga. Geogr. J. 74: 109–132.

13. THE WHITE NILE FROM MALAKAL
TO KHARTOUM

by

J. RZÓSKA

After leaving the swamps the White Nile flows in two sharp turns into its final stretch to Khartoum. South of Malakal it receives its last but most important tributary, the Sobat. With an annual discharge of 13 × 10^9m^3, the Sobat compensates for the water losses in the swamps, and completes the total water budget of the White Nile (see ch. 8) and its drainage network (see Fig. 14). The western affluents from the Zaire (Congo) divide contribute little, being mostly lost in the Ghazal swamps.

The Sobat derives its water supply mainly from the southern Ethiopian Plateau and therefore differs in its regime from the Bahr el Gebel. Its discharge is highest from May to October and has a stowing back effect on the swamp rivers. The Sobat carries a fine mineral (volcanic) sediment of whitish colour which persists in the White Nile downstream and may be one of the reasons for the colour difference between the White and Blue Nile. A well developed plankton appeared in its end run in 1953–1956. There are considerable swamps in the Sudanese part of the Sobat course and its tributaries, now choked with the water hyacinth (see ch. 21).

Now the White Nile flows north into an increasingly arid landscape of dry savanna (see Figs. 10 and 11). At Malakal there are six rainless months, at Kosti nine months without rain and this trend increases towards the north. This river stretch, about 800 km. long, differs also geomorphologically from the southern parts. Berry (ch. 2) describes it as 'a most unusual river with a gradient lower than that of the Sudd'; the gradient is only 11.60 m in 800 km., 1:69743. This results in a lower current velocity of approximately 0.5 m/sec at Malakal to below 0.1 m/sec near the Gebel Aulia Dam, 44 km south of Khartoum, when the dam is open. For comparison, current velocities were 1.10 m/sec at Bor and 0.67 m/sec at Lake No, a stretch of about 770 km.

According to Berry the White Nile north of Malakal flows in a 'well defined channel or channels within a broad valley', suggesting an 'old and fundamental drainage line'. Although braided there are few meander curves. The depth rarely exceeds 10 m. The shores are flat on the western side, higher on the eastern side where the Gezira alluvial was created in the past (see chapter 4c.).

Before the erection of the Gebel Aulia Dam in 1936, the braiding seems to have been more extensive. Numerous islands were more conspicuous, acacia forests fringed large parts of the river valley and large inundations occurred during the moderate high water (August to January). These inundations were strongly influenced by the ponding effect of the powerful Blue Nile flood; this effect still remains in a lesser way now. But in the past it took on enormous dimensions as shown by the remarkable lake formation which existed from about 11,800 to 7000 years BP (see chapters 2 and 4a.). The ponding effect has been artificially increased in 1936 by the Gebel Aulia Dam (Fig. 43) which acts as regulator, keeping a steady flow of the main northern river, when the Blue Nile is low. The dam, owned by Egypt, stores $3.5 \cdot 10^9$ m³ from July to February, with a gradual discharge ending in June when river conditions return. When full the influence is felt upstream for 500 km. The stowing effect of the dam causes water level changes: at Malakal 2.25 m, Kosti 4 m, near the dam 6 m.

With the sequence of management from opening to closure of the dam, river conditions are followed by a 'flowing lake' state. Currents fall upstream and so does their capacity of carrying a sediment suspension. The water clears gradually downstream, photosynthesis sets in and, along with the increase of algal densities, zooplankton appears in a pure form. This plankton suspension persists downstream and it was a surprising discovery for the biologists who began to work at the University College in Khartoum in 1947. At that time this natural and now predictable sequence was a 'discovery' in view of the lack of information for any long tropical river.

Fig. 43. The Gebel Aulia Dam on the White Nile south of Khartoum; Note the bare surrounding landscape; Photo Capt. Lovelock.

216

Starting from observations at Khartoum, the biologists of the newly created University College began to study, from 1947 onwards, the composition (Brook 1954) and the seasonal appearance (Rzóska, Brook & Prowse 1955, Prowse & Talling 1958). Later, understandably, the origin and spatial distribution of plankton was studied in longitudinal surveys (Brook & Rzóska 1954, and unpublished data) and was followed by investigations on the longitudinal succession of water characteristics (Talling 1957); finer problems of primary productivity and detailed densities of plankton followed. Talling expanded his work later to fundamental studies of primary production in Africa and compared these with results in Europe.

Years later in 1965 D. Hammerton took up further studies after the *Eichhornia* invasion (see ch. 21). His results on primary production are unpublished but have been made available by courtesy. He followed the rate of primary production at a site in the Gebel Aulia basin during a whole day, and the longitudinal development at six stations from Malakal to the Gebel Aulia dam. The results of all these investigations will be presented in the General chapters 24 to 27 because rivers, though-out ecosystems in the accepted sense, are in some respects coherent 'systems' even over thousands of km. Generally, one can say that limnology in the Sudan started with the White Nile work.

Biomass assessments of zooplankton and benthos in the White Nile have been carried out by Monakov (1969) and physical and chemical features of Nile water by Kurdin (1968), both members of a Soviet expedition in 1963/64. The results will also be discussed in the general chapters. Monakov's observations are the only direct benthos data for the Sudan; but a number of other studies especially on insect emergences exist and are incorporated in chapters 23 and 24.

Finally, another look at the general character of the river from Malakal northward. Although the big swamps end before Malakal, the river is fringed by riverain reed belts, varying in width but gradually thinning out towards the north. P. Gay has studied the composition and zonation of these wetlands in 1956/7. He took several hundreds of photographs, which allowed him to pinpoint the arrival of *Eichhornia* in 1958. He published two interim accounts on the 'Riverain Flora of the White Nile and the Bahr el Ghazal' in the Annual Reports of the Hydrobiological Research Unit (University of Khartoum) no. 4 (1956/57: 10–20 and no. 5 (1957/58): 7–15. Papyrus finds its northern most outpost at Kosti, where single specimens and small stands are met first in its retreat from the north. Most of the plants are (were) transported by floating islets stranding along the river and ending their existence usually at Kosti (for conditions after the Eichhornia invasion, see ch. 21).

The biological effects of the dam were clearly reflected in fisheries; camps sprung up during the 'flowing lake' phase and the considerable catch was either salted or brought to the Khartoum market. This effect

217

of dam reservoirs is now seen on much larger scale in Lake Nubia (see ch. 18 and 19).

The final stretch of the White Nile flows not only through a different vegetational zone but also through a zone of a different human environment. Sudanese Arab tribes occupy a broad belt across the country. This is the southern limit of Arab-Muslim penetration in this part of Africa. They were halted in their further advance by insect pests, mainly Tabanids, vectors of camel and horse diseases (see ch. 23).

REFERENCES

These will be incorporated in ch. 24–27.

THE BLUE NILE SYSTEM

by

J. RZÓSKA

The Blue Nile descends from the Ethiopian high plateau (Fig. 44). It receives a great number of tributaries in its upper course in Ethiopia and two further, the Dinder and Rahad, in its lower, Sudanese, course. Although the total drainage area is relatively small, 324,000 km² (Hurst 1950), it supplies 58% (excl. losses) of the total water of the Nile system and almost all the sediment which has built up the alluvial river valley and the Delta in Egypt. The Blue Nile differs from the White Nile in two main, spectacular, characters: The enormous seasonal flood with a discharge 40–70 times that of its low season and a load of sediments of specific mineral composition amounting to 100. 10⁶ metric tons per year.

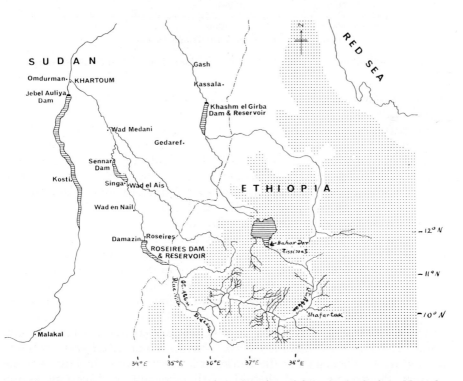

Fig. 44. Map of the Blue Nile system and the location of dams in the Sudan; Note the numerous affluents in the semi circular sweep of the Gorge and location of some sites; Dotted areas are over 1500 m a.s.l.; adapted from Min. of Irrigation and other sources.

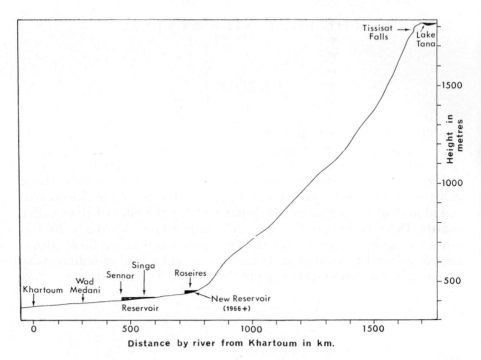

Fig. 45. Longitudinal profile of the Blue Nile; adapted from Talling & Rzóska 1967.

Further differences exist in the chemical composition of the water and the incised valley.

Three landscapes encompass the river in its course. The cool High Plateau with a montane vegetation greatly altered by man surrounds Lake Tana and its 60 affluents, of which the Little Abbai is the most important. Quite different is the next sector, about 650 km long, where the river cuts through the mountains in a gorge. In some places this is 1,000 m deep and descends steeply to the plains (Fig. 45). This is a violent landscape largely untouched by man.

The Sudan plains, an Acacia-grass savanna, are increasingly arid towards the north and the river meanders through them with a more gentle slope, falling from about 500 to 374 m in its altitude in its course of 900 km. This section has undergone great changes in water regime and its biology due to the construction of two dams, at Sennar in 1925 and Roseires in 1966.

The origin and history of the Nile system are treated in ch. 1–3; only some brief notes are included here. The L. Tana basin was probably formed in the Pliocene and the Gorge excavated during the late Pliocene to early Pleistocene (see Epilogue). Earlier volcanic and tectonic action created the uplift of the Ethiopian Highlands. Before the present course of the Blue Nile was established, probably about 30,000 years BP, older

220

channels have probably existed along the western flanks of the Plateau and the southern Red Sea hills. Distinctive Blue Nile or Atbara sediments have appeared in the ancient Nile in Nubia and Egypt on some early occasions. A conspicuous creation of the Blue Nile in the past was the Gezira plain, now the most important agricultural region of the Sudan.

14. LAKE TANA, HEADWATERS OF THE BLUE NILE

by

J. RZÓSKA

The Little Abbai, one of the 60 affluents of the lake is usually regarded as the source of the Great Abbai (Blue Nile). It springs lie about 100 km south of the lake. The intrepid Scottish traveller James Bruce saw these springs in 1771 and derided with scorn (1790) their discovery by Portuguese Jesuits, Fr. Pedro Paez in 1613 and later Hieronimo Lobo (B. Tellez 1710, Travels of the Jesuits in Ethiopia). But Fr. Lobo has left us his 'Breve relacao do Rio Nile' which, translated by Sir Peter Wyche, was published in London in 1669.

On the 22 October 1668 a communication was read to the Royal Society in London by the Rev. Dr. R. Southwell under the title '*An extract of several relations delivered by P. Hiernonymo Lobo, a Jesuite, and an eye witnesse of most ye particulars contained therein and by same imparted to Dr. R. Southwell, who generously imparted it for publication*' (R.S. Classified Papers 1660–1740 VII, 1). In this paper Fr. Lobo interpreted correctly the origin of the Blue Nile, and that of the flood and the inundation of Egypt, as a result of rainfall in Ethiopia.

The hydrological exploration of this region on a scientific level had to wait until the beginning of the XX century, when the Sudan became again accessible to the outside, after the political interruption by the Mahdi's revolution. Under the pressure of agricultural needs in Egypt several expeditions of the Egyptian Irrigation Department, later Ministry of Public Works, e.g. Grabham & Black in 1920 (1925), established the main outlines of the water regime. Previously and simultaneously Lake Tana was visited and mapped by explorers. Morandini (1940) has collated the results of these efforts and reproduced a series of earlier maps. He was a member of the Italian 'Missione di Studio al Lago Tana' which worked in Ethiopia from 1937 onwards and produced numerous studies. A large amount of hydrological and geographical data are contained in the volumes of 'The Nile Basin' e.g. vol. VIII by Hurst (1950). The same author wrote also the well known monograph on the whole riversystem (Hurst 1952). Cheesman (1936) was one of the few who described the Blue Nile Gorge; there are many other smaller contributions.

The origin and geology of the Lake Tana basin has been briefly mentioned above and in more detail in chapter 2. – The drainage basin of the lake occupies about 16,500 km², of which only 3,156 km² are occupied by the lake itself at present. Terraces, differently interpreted,

223

point to a much larger previous lake body; Nilsson (1940) recognises these at 100 m above the present level, Minucci (1938 quoted after Morandini 1940) only at 10 m. The lake's altitude is 1829 m, its maximum length 78 km, width 67 km, and the shoreline is 385 km with three groups of islands.

The maximum depth is 14 m, mean depth 8.91 m. The bathymetric map (Fig. 46) shows a gently sloping basin, a space photo (Fig. 47) shows the accuracy of the surveyors. The shores are partly rocky but especially

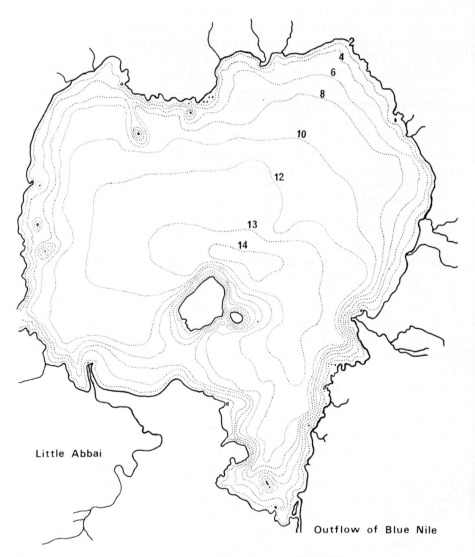

Fig. 46. Lake Tana, bathymetric map; from S. Morandini 1940.

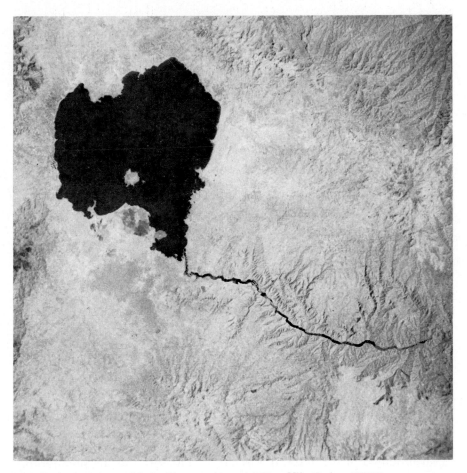

Fig. 47. Space photo of Lake Tana and part of Blue Nile Gorge; ERTS.

at the mouth of affluent rivers there is a littoral flora, including papyrus *(Cyperus papyrus)* from which boats are made, resembling those of ancient Egypt (Fig. 48). The bottom is covered largely with soft sediments; pollen analysis was recently undertaken as mentioned by Livingstone in chapter 3.

The climate is fluctuating between extremes; temperatures at daytime vary from 23 to 30 °C, falling during the night to 6–8 °C (Morandini 1940 p. 126, Hurst 1952 p. 170).

Rainfall may reach up to 2000 mm per year in the surrounding plateau; 90% of this falls in one rainy season from June to October with a peak in July and August. It is almost rainless in the winter. Relative humidity is above 80% during the rains and little more than 30% in the dry season (Hurst 1952).

225

Fig. 48. A reed boat, 'tankwa', on Lake Tana; this type of boat is frequently depicted on ancient Egyptian murals, this shows its wide spread of use; Photo J. F. Talling.

Limnology

The water budget of Lake Tana fluctuates with the drying out of many affluents in winter and the big rainfall in summer. By May–June lake levels may fall about 1.90 m reducing the outflow at Bahar Dar drastically as documented in the graph of discharge (Fig. 49) and the photographs of the Tississat Falls, about 30 km below the outflow (Fig. 50, 51). Evaporation, according to Grabham & Black (1925) quoted by Morandini (1940) varies from 1.5 to 6 mm per day; a white band of dried up algae appears on rocks during low water. According to the Italian hydrologists (Morandini l.c., after Gentilini 1936) the lake has a buffering effect on the flow of the Blue Nile at the outflow.

Thermics

Only conditions during the winter months, hot and dry, have been recorded by Morandini. In contrast to air temperatures, those of the water fluctuate only between 21 and 25 °C. Talling measured in March 1964 a range of 23.3–25.3 °C. There is only a small gradient to the bottom, but thermoclines of short duration and very variable may be formed at depths of 1–4 m, with a temperature gradient of only 1 °C.

Transparency by Secchi disc was low and varied between 0.80 and

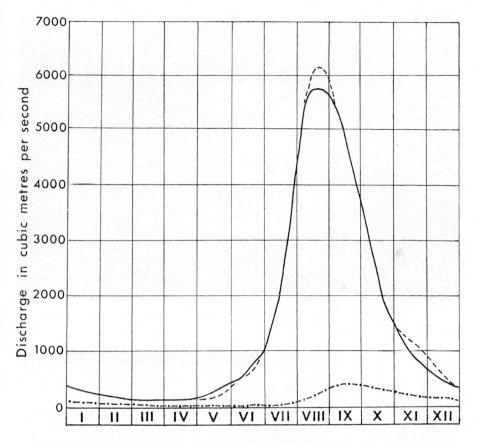

Fig. 49. The discharge of the Blue Nile at Roseires 1912–1956; Note the small contribution of Lake Tana -.-.; additions to the solid line curve represent variations; adapted from Min. of Irrigation, Sudan and S. Morandini 1940.

1.30 m. The colour of the water was greenish-yellow. Later, in 1964, Talling (unpublished) obtained photo-electric measurements of light penetration near Bahar Dar. These indicated a euphotic zone of approx. 3 m, with maximum transmission in the red spectral region. Seiches occur of 130–210 minutes duration, as observed by Grabham & Black (1925) and quoted by Morandini. Some currents were said to reflect the influence of certain affluents. Water chemistry will be treated by Talling in ch. 26.

Hydrobiology

Brunelli & Cannicci (1940) described the shore flora and fauna and the plankton; Bini (1940) dealt with the fishes. These investigations were made in 1936/7 and reflect the opinions prevalent at that time; according to

227

Fig. 50. The Tississat Falls near the outlet of the Blue Nile from Lake Tana during the rainy season (August-September).

Fig. 51. The Tississat Falls during the dry season (May); Photo J. Morris.

the classification of Thienemann & Naumann the lake is regarded as 'oligotrophic'.

The shore flora consists of localised stands of: *Nymphaea coerulea* Sav., *N. lotus* L., *Rotala repens* Koehne, *Ceratophyllum demersum* L., *Vallisneria*

228

spiralis L., *Potamogeton schweinfurthii* A. Benn., *Cyperus papyrus* L., *Typha domingensis* Pers., *Oenanthe palustris* (Chiov.) Norman, *Hydrocotyle ranunculoides* L.F., *Cyperus* sp.; *Pistia stratiotes* L. In swampy areas *Echinochloa stagnina* (Retz.) P. Beauv., *E. pyramidalis* (Lam.) Hitchc. & Chase, *Panicum subalbidum* Kunth, *Polygonum barbatum* L. and *P. senegalense* Meisn. have been recorded by the authors. (This list has been corrected by Dr. G. Wickens, Kew, Ed.).

PHYTOPLANKTON

In the large list of phytoplankton there is a low-representation of Cyanophyceae; the similarity of the algal composition to that of other African lakes is noted. There are, for example, large and apparently planktonic species of the diatom *Surirella*, also noted by Talling (unpublished) in 1964. But according to Talling (Talling & Rzóska 1967) the exact determination by the Italians of some species and their role in the plankton is open to doubt. This seems especially applicable to the species of the diatom *Melosira*; samples collected by Talling in 1964 show a plankton dominated by *Melosira granulata* var. *jonensis f. procera*. The species recorded by the Italians are not those prevalent in the Blue Nile at Khartoum as determined by Brook (1954) and by Prowse & Talling (1958) see ch. 27. No quantitative samples for counts of organisms have been collected from L. Tana and no estimates of primary production are available so far. A single estimation of chlorophyll *a* content, made by Talling (unpublished) in 1964 near Bahar Dar, was relatively low —3.7 mg/m³.

ZOOPLANKTON

This shows the usual assembly of circumtropical, cosmopolitan species and some African forms. Noteworthy is the presence of *Daphnia longispina* which appears in the Nile system in both headwater lakes and otherwise only at the lower end of the Nile.

Daphnia barbata, an African species which is present in the river from Roseires onwards was not found. Net samples collected by Talling in 1964 showed a different numerical composition than those collected by the Italians and contained *Daphnia lumholtzi* not noted by them (Talling & Rzóska 1967). Such changes in composition with time are, however, common in water bodies everywhere; the Rotifers present the usual almost cosmopolitan assembly. No quantitative sampling has been done by Brunnelli & Cannicci. A number of good microphotographs are included in their paper.

Of the other groups the Mollusca are important for zoogeographical reasons; a preliminary list was given by Piersanti (1940) but has been corrected and revised by Bacci (1951/52). This seems to be the last critical appraisal of the mollusk fauna; 15 species live in L. Tana: *Theodoxus africanus* (Reeve), *Bellamya unicolor unicolor* (Oliv), *B. unicolor abyssinica* (Martens). *Melania* (Melanoides) *tuberculata* (Müll), *Radix pereger* (Müll), *R. caillaudi caillaudi* (Bourg), *Bulinus hemprichi sericinus* (Jickeli), *Biomphalaria rüppelli* (Dunker), *Unio elongatulus dembeae* (Sowerby), *Unio abyssinicus* Martens, *Apatharia rubens caillaudi* (Martens), *Mutela nilotica* (Caillaud), *Etheria elliptica* Lamarck, *Corbicula fluminalis consobrina* (Caillaud), *Byssanodonta parasitica* (Deshayes).

Of the 15 species, 4 are of palaearctic character: *Unio abyssinicus* is localised in L. Tana and some related forms exist now in Syria and Palestine, fossil specimens have been found in the Nubian Nile and in the Faiyum (see ch. 4a.); the other *Unio* belongs to a group of forms with wide distribution, found fossil in L. Rudolph. Both are not found at present in the Nile and interesting conclusions on past climatic changes are mentioned in ch. 4a. The other palearctic species are *Radix pereger* and *Theodoxus africanus*. All other species are nilotic; *Bellamyia unicolor abyssinica* is endemic. A negative character of the Tana mollusc-fauna is the absence of the genera *Cleopatra*, *Pila* and *Lanistes*, which are frequent in the Nile and elsewhere in Africa. The palaearctic elements in L. Tana seems to have been distributed widely in the Nile in the later Pleistocene but have disappeared, presumably by warmer water temperatures during the onset of the present desert phase (see 4a.). Bacci (l.c.) decribes the adjacent but separated fauna of Eritrea; 25 species of mollusks are present of which only 6 are common with L. Tana; 6 are palaearctic, the other African or specifically nilotic, a zoogeographically interesting composition. Bacci makes some remarks on lakes in other parts of the country, the Omo region and Lake Rudolf of considerable interest to the present and past zoogeography of this part of Africa. – Amongst the snails are a number of vectors of Bilharzia and the disease is rife.

FISHES

The early work by Bini (1940) has now been superseded by recent results. Greenwood (ch. 7) refers to the Tana fauna as an 'extremely truncated' Nile fauna with little endemism; only 8 or 9 species are now recognised with one outstanding feature of a Euro-Asiatic genus present.

There are no crocodiles in Lake Tana though present in the Gorge; the hippopotamus has vanished.

The general zoogeography of the Ethiopian High Plateau differs from that of the rest of the Nile basin; some remarks will be found in the chapter on the Blue Nile Gorge by P. Morris. A number of papers have appeared on the freshwater fauna of Ethiopia, but none of these refers to L. Tana or the Nile.

Descent to the plains, the Blue Nile Gorge

At Bahar/Dar the Great Abbai (Blue Nile) leaves L. Tana and passes through 30 km of marshes, meadows and cataracts to the Falls of Tississat 50 m down (see Fig. 50), and enters into the Gorge. Numerous tributaries from the surrounding massifs enter the great arc-like sweep of the river through deeply incised ravines representing extensive excavations and indicating a long period of erosion and higher rainfall in the past (see (ch. 2 and the new age dating, mentioned in the Epilogue).

The hydrology of the Gorge is as yet little known but according to Morandini (1940 quoting Visentini) two thirds of the whole supply of the Nile flood is derived from the lower part especially from the Jamma, Guder and Didessa rivers. Many of the smaller tributaries are temporary; Hurst (1950) observed the effects of a thunderstorm on one of these; a mere trickle of water changed within few minutes into a spate 2–3 m deep and subsiding within few hours. Morris in the next pages describes his experiences with such spates. A graph (see Fig. 39) illustrates the annual course of discharge at Roseires and the contribution to it of Lake Tana and the rivers in the Gorge. For the total contribution of the Blue Nile at Khartoum see ch. 8.

The exploration of the Gorge has had a short but eventful history with several attempts to ascend or descent it from 1902 onwards. Most of these were unsuccessful and did not contribute to our knowledge. In 1926 R. E. Cheesman, a British consul in Abyssinia, surveyed the Gorge on foot and mule. Forced by the enormous difficulties of the terrain he had to confine much of his work to the rim of the precipitous valley. Nevertheless he left a very valuable record, including many zoological observations (Cheesman 1936). The contribution which follows is based on the first successful descent of the whole Gorge in 1968, by a team of military and scientific personnel under the technical direction of Major Blashford-Snell (1970). Previous expeditions of the group in 1964 and 1966 did much to prepare for the successful work on this last unexplored region of the Nile system. P. Morris who participated as zoologist describes in the following pages the chief characteristics of the Gorge and its riverain zoology. Unfortunately no botanist nor hydrobiologist participated.

REFERENCES

Bacci, G. 1951–1952. Elementi per una malacofauna del'Abissynia e delle Somalia. Ann. Mus. Civico di Storia Nat. Giacomo Doria 65: 1–44.

Bini, G. 1940. I pesci del Lago Tana. In: Brunelli & Cannicci.

Bruce, J. 1790. Travels between 1768–1773 to discover the source of the Nile. vol. 1–5. Edinburgh.

Brunelli, G. & Cannicci, G. 1940. Le caratteristiche biologiche del Lago Tana. Missione di Studio al Lago Tana. Ricerche limnologiche B Chimica e biologia. vol. III (2): 71–116. Reale Acad. d'Ital. Centro Studi Afr. Orient. Ital.

Cheesman, R. E. 1936. Lake Tana and the Blue Nile. Mac Millan, London.

Grabham, G. W. & Black, R. P. 1925. Report of the Mission to L. Tana, 1920–21. Min. Publ. Works Egypt, Govt. Press, Cairo.

Hurst, H. E. 1950. The hydrology of the Sobat and the White Nile and the topography of the Blue Nile and Atbara. The Nile Basin, vol. III. Min. Publ. Works, Egypt. Govt. Press, Cairo.

Hurst, H. E. 1952. The Nile, a general account of the river and the utilization of its waters. Constable, London.

Missione di Studio al Lago Tana. (Numerous contributions from 1938 onwards, Reale Acad. d'Italia).

Morandini, G. 1940. Ricerche limnologiche, Geografia-Fisica, vol. III, 1. Missione di Studio; pp. 319.

Nilsson, E. 1940. Ancient changes of climate in Brit. East Africa and Abyssinia. In: A study of ancient lakes and glaciers. Geografiska Annaler 22: 1–79.

Piersanti, C. 1940. Molluschi del L. Tana della zona finitima. Missione di Studio al Lago Tana. vol. III, 2: 233–263.

Tellez, B. 1710. The travels of the Jesuits in Ethiopia. London.

Talling, J. F. & Rzóska, J. 1967. The development of plankton in relation to hydrological regime in the Blue Nile. J. Ecol. 55: 637–662.

15. NOTES ON THE BIOGEOGRAPHY OF THE BLUE NILE (GREAT ABBAI) GORGE IN ETHIOPIA

by

P. MORRIS, M. J. LARGEN & D. W. YALDEN

The authors visited the Gorge in August–September 1968 on the 'Great Abbai Expedition', the organisation of which was described by Blashford Snell (1970) and Snailham (1970). The river was explored from Shafartak road bridge downstream to Sirba (about 400 km), using four metal boats. The journey took one month and was punctuated by stops of 1–5 days at 'Forward Bases' along the river. General collecting of vertebrates was done at these, and it is this part of the river with which we are most familiar.

The second phase of the expedition began at Bahar Dar, where the Nile leaves Lake Tana. Rubber boats took a small team downstream to Shafartak, supported by supply parties moving overland. The zoological group did not travel this part of the river, though did spend a few days at Tisisiat Falls and two weeks collecting elsewhere in Ethiopia. Specimens were brought to the British Museum (Nat. Hist.) in London and some are deposited with the Haile Selassie University in Addis Ababa.

During the whole expedition collecting was carried out in four major climatic regions: the high cool plateau (Debre Marcos & Ghimbi); the wet grassland of Bahar Dar; the Nile Gorge and the arid lowlands of Eastern Ethiopia.

The Highlands are separated by their altitude from the lower parts of Africa all round and consequently harbour a large number of endemic animals. The Gorge is a unique habitat, open to lowland animals only at the western end, being hemmed in by mountains elsewhere.

The main collecting stations in the Nile Gorge were from Shafartak Bridge (38° 17′ E 10° 03′ N) westwards, ie downstream, to Sirba (35° 30′ E 10° 05′ N) and were all at approximately 1,000 m altitude.

Forward Base 1 – Confluence of Guder R. & Nile (37° 48′ E, 09° 30′ N).
Forward Base 2 – 10 km west of Mabil (36° 45′ E 10° 19′ N).
Forward Base 3 – Confluence of Didessa R. & Nile (35° 38′ E 10° 05′ N).
There were also other minor collecting stations in between these.

Nile Gorge Physical Factors

The Gorge is sometimes several kilometres wide, the sides descending in steps. At the bottom of this chasm the river flows for a considerable distance in a narrow 'inner gorge'; with 100–150 metre high cliffs often rising sheer from the water course, especially for the first 100 km downstream of Shafartak and downstream of Tisisat.

Most of the Gorge, especially around Forward Bases 1 and 2 has very steep slopes which rise immediately from the river bank. Hillsides are well drained with light soil and much bare rock and loose scree. Sandbanks occur at the river's edge. In places, a narrow fringe of forest at the river's edge remains green all year round, (therefore forming important shelter in the dry season), but often there is little undergrowth below these trees. Elsewhere the very dense undergrowth of thorn bushes and shoulder-high grass is often quite impenetrable, the only access usually being by way of dry stream beds and gulleys. (Fig. 52).

The Gorge seems to generate its own climate. Large quantities of bare rock store up heat during the day so that the Gorge is very noticeably warmer at night than the plateau above and more humid. Often, especially in the early morning, clouds form in the Gorge whilst the high ground is clear. It is possible to stand on the plateau above Shafartak and look down on the clouds in the Gorge below.

Towards Forward Base 3 and Sirba the river banks are flatter, sandy and low, with gently sloping hills set back from the river. Along most of the Gorge there is no cultivation and the land is uninhabited close to the river until west of Forward Base 2. Small infrequent plots of maize

Fig. 52. The Blue Nile Gorge near Shafartak during the dry season; Note clear and shallow water and dense growth on river bank; Photo J. Morris.

234

cultivation appear after Forward Base 2; with permanent but isolated settlements from Forward Base 3 onwards.

There is a major contrast between the dense vegetation, steep rocky slopes, well drained soils and hot, humid air of the Gorge and the open, flat, wet, cold plateau immediately adjacent to it but 1000 metres or more above it.

In August and September, the high plateau above the Gorge is very cold at night, and wet. It is extensively cultivated and bears little vegetation more than knee high owing to overgrazing and collection for fires. Around Bahar Dar and Tisisat, fields were under 30 cms of water during our visit, forming large pools and marshy areas dotted with boulders and isolated clumps of bushes. At Tisisat itself, constant spray maintains a very localised patch of thick riverine forest adjacent to the Falls.

It should be stressed that the descriptive notes given for the collecting localities apply to the time of our visit (Aug–Sept incl.), towards the end of the rainy season. Consequently lush vegetation, roadside pools and fast flowing streams are described. The river was in full flood and flowing very fast. The water was thick with suspended mud and usually very turbulent in the main river course, though relatively static backwaters occurred, especially at the confluence with tributaries. During our visit, the Nile was sometimes 100 metres or more wide. The river drains a vast upland catchment area, so that in the narrower parts of the river, the water level rose rapidly following storms – over 3 metres in a single night at Forward Base 1.

When rainfall is reduced or absent, the countryside may undergo great changes, particularly in the Nile Gorge. During the dry season, (especially May), aerial pictures and reports from a reconnaissance party and other travellers show that the Gorge is very arid and hot. Bushes lose leaves exposing bare rocky ground. Only the fringe of trees beside the river, with roots reaching to the water table, remain green and in leaf. The river almost ceases to flow and meanders between broad shingle banks. The water is often only about 50 cms deep (easy to ford) and in places rocky sills interrupt the flow causing it to form a long series of static, hot, shallow lagoons. (For dramatic demonstration of aridity of dry season, compare photographs of river flow at Tisisat Falls in May & August, Fig. 50, 51).

This seasonal contrast cannot fail to have a very severe limiting effect on the fauna. A general discussion follows; details of the animals collected appear in Hill & Morris (1971) and Corbet & Yalden (1972). Reference may also be made to the bird report by Johnson (1972).

The Fauna of the Blue Nile (Great Abbai) Gorge in Ethiopia

Despite its diversity and potential interest, the fauna of Ethiopia has been poorly studied. The Great Abbai Expedition found at least a dozen species of vertebrates, previously unknown in Ethiopia (including two species new to Science). Many of these were fairly abundant animals, indicating just how poor previous collections had been. Against a background of uncertainty and very incomplete knowledge of animal distribution and ecology within Ethiopia as a whole, an analysis of the Nile Gorge fauna must be superficial and tentative. A number of general points are nevertheless evident:

1. In the wet season the river itself constitutes a very inhospitable environment for aquatic life. The powerful flow of sediment-laden water (the river is chocolate brown and even a cupful would yield a measureable amount of solids if allowed to settle), had scoured rocks and near Forward Base 1 had polished them so that they shone in the sun as though permanently wet. Daily fluctuations in the river level must aggravate the situation.

The powerful flow is in places as fast as that of a small mountain stream and might be expected to flush out most aquatic organisms. Certainly macrophytes were only seen in relatively static areas out of the main stream and in tributaries.

The steepness of the Gorge sides meant that wet season flow in tributary streams would be very fast and turbulent. Many small streams appeared to flow only during storms; at other times forming pools and damp rocky areas. These provided a haven for small fish (*Barbus* sp.) amphibian larvae and the side necked terrapin *(Pelomedusa subrufa)*.

2. In the dry season the river almost ceases to flow. In places it breaks up into a series of shallow, static pools or meanders between wide shingle banks. Elsewhere it is interrupted by massive boulders and wide sills of bare rock; all of which are under turbulent water in the wet season.

The shallow, hot pools cannot provide a very secure habitat, especially since they are liable to dry up altogether or be inundated when the rains come.

Insects (especially mosquitos) were noticeably scarce during the wet season, but the few previous travellers visiting the Gorge at other times of the year have reported plagues of them. Presumably the principal aquatic insects are those with short life cycles that can develop during the dry periods in great numbers, be flushed out by the rains, and later swiftly recolonise the river from small reservoir populations in backwaters. There is little prospect of permanent plankton populations being established along considerable stretches of the river[1].

The apparent general scarcity of fish found by the Great Abbai Expedition in the upper parts of the Nile could be attributed to ineffectual

[1] EDITOR. Talling once found rudiments of phytoplankton especially *Cyclotella* in the river water at the Tississat Falls and at the Shafartak bridge (Talling & Rzóska 1967 and personal communications).

fishing methods, but the same procedures were quite successful when used elsewhere in Ethiopia. More probably the scarcity was a genuine one caused by the extreme force of the currents and muddiness of the water. Food for fish must be in short supply, the currents sweeping away any small prey or suspended matter, though during the dry season there is probably an abundance of insect larvae in the shallow river water. In relatively few places is the river calm enough to allow growth of aquatic vegetation.

Another indication of fish scarcity was the small number of fish predators seen. A few egrets, herons, fish eagles and crocodiles were observed, but these were relatively infrequent compared with their abundance in waterside habitats elsewhere in East Africa. Very few other common riverside species of birds were seen either, suggestive of food shortage.

Among the few fish that were found in the Gorge, catfish (*Synodontis, Bagrus* & *Heterobranchus*) predominated. With their long, sensitive barbles, these are well adapted to life in turbid water and the electric fish found (*Melapterurus electricus* and *Mormyrops anguilloides*) also have the advantage of possessing a non-visual navigation sense. Fish with well developed eyes were not found in the muddy Nile until several *Eutropius niloticus* were caught in the clearer water at the mouth of the Didessa River. A few small *Barbus* sp. were collected in the Gorge, but from a clear pool fed by a stream, not the main river.

3. Despite its equable climate and lush appearance in the wet season, the Gorge is a harsh, hot place at other times. The fauna, as observed during the wet season, must therefore be capable of migrating out of the Gorge, or withstanding semi-desert conditions for much of the year. Examples of the latter include scorpions, sun spiders (Solpugida) Agamid lizards and spiny mice *(Acomys)*, animals normally associated with arid areas rather than the relatively lush habitats in which they were found by the Expedition.

Migration in such rocky and steeply sloping terrain would obviously be difficult and may account for the few sightings made of the larger mammals or their tracks and droppings. Klipspringers *(O. oreotragus)* and monkeys *(Cercopithecus aethiops)* were sometimes observed: baboons *(Papio anubis)* were the only larger mammals that were at all abundant.

Bats are more mobile and therefore more likely to migrate in and out of the Gorge seasonally. In fact the bats of the Gorge were remarkably abundant and diverse. 14 species (+ one whose provenance is uncertain) were found in the Gorge, some in considerable numbers, compared with three at Ghimbi on the plateau. The Gorge bats included at least four species *(Taphozous perforatus, Nycteris thebaica, Hipposideros caffer, Triaenops persicus)*, which are known to occur in hot dry areas elsewhere in Ethiopia and may well survive the dry season feeding on insects without needing to emigrate. They are also species known to roost in caves and rock crevices where they would be protected from daytime heat.

237

The Megachiroptera collected *(Epomophorus anurus, Micropteropus pusillus)* would probably not be able to find food in the dry season, when most of the vegetation is dry or dead. In the wet season they were found living in, and feeding on the fig trees fringing the Nile. These bats might therefore be expected to migrate, a relatively simple feat since fruit bats are strong fliers. Indeed their principal home may be far away and they make only temporary use of the Gorge as food becomes seasonally available there. Since fruit bats are not common on the high Ethiopian plateau we may suppose that any emigration undertaken would be along the river (presumably westwards to lower altitudes) rather than laterally (North or South) to the Ethiopian plateau.

4. The Gorge provides a habitat which is quite different and separate from the highlands (ca. 2000 + metres above sea level), adjacent to it. The Gorge traps heat and is at a lower elevation; both factors make it consistently and noticeably warmer than the plateau at all times of the year, but most notably at night in the wet season. The contrast between the Gorge and the Highlands is evident from the general nature of the vegetation and in the fauna of the two areas.

The Great Abbai Expedition collected at two major highland sites: Debre Marcos (c. 2,500 m) and Ghimbi (c. 2100 m), respectively 40 km North and 100 km South of the Nile. Collections here may be compared

Table 1. A selection of species collected by the GAE in the Nile Gorge or nearby on the Ethiopian plateau. (Species with respectably sized samples only).

AMPHIBIA	Gorge (ca 1000 m)	Plateau (Ghimbi &/or Debre Marcos) (2000 + m)	Bahar Dar (1850 m)
Bufo kerinyagae	0	22	0
Leptopelis sp.	0	1 (+ 12 at Enjiabara, 2700 m)	0
Leptopelis bocagei	1	0	18
Ptychadena erlangeri	0	76	2
Ptychadena huguettae	8	0	0
Phrynobatrachus natalensis	3	9	17
REPTILES			
Pseudoboodon lemniscatus	0	10	0
Typhlops spp.	0	2	1
Varanus niloticus	2 (many seen)	0	0
Mabuya striata	0	19	0
M.quinquetaeniata	14	8	1
M.isselli	3	8	0
Latastia longicaudata	5	0	0

In addition to species in Table, two other Amphibians, plus seven reptiles were collected in the Gorge. Many additional species were taken in the highlands and at Bahar Dar.

with those made at river level in the Gorge (c. 1000 m) and at Bahar Dar (c. 1800 m) where the Nile leaves Lake Tana. Collections were somewhat random, often not made by zoologists and frequently species are represented by only single specimens; so conclusions again must be somewhat tentative.

It was noticeable that many species were found either in the Gorge or on the plateau, not both, despite the nearness of collecting localities in terms of horizontal distance (Table 1).

Species endemic to Ethiopia might be considered as a speciality of the highlands, but the endemic lizard, *Mabuya isselli*, was found in the highlands and also in the Gorge. The chameleon, *Chameleo affinis*, is another endemic find on the plateau, but our failure to collect it in the Gorge may have been due to the fact that specimens were procured by local boys on the Plateau, whose talents were not available on the river. A similar collecting bias affected the snakes: many were obtained from local people at towns on the plateau, but their apparent absence from the Gorge (only one collected) may have been real or may have been due to the lack of local assistance in finding them.

The grass frog, *Ptychadena erlangeri*, was particularly abundant on the plateau, but was not encountered in the Gorge.

A more instructive comparison is made by considering what was found in the Gorge, but not encountered in the highlands. Most of these species are essentially lowland forms widely distributed in the savanna areas of Africa. They include the one snake collected in the Gorge, the spitting Cobra *(Naja nigricollis)*, which is common in East Africa, and the gecko, *Hemidactylus brooki*. The lizard, *Latastia longicaudata*, is widespread from Nigeria to Somalia and Tanzania, but was collected by the Great Abbai Expedition only in the Gorge, near the River Mugher. *Rana occipitalis* is a common frog in the savanna of East and West Africa. It was also collected at Sirba, close to the Nile, though not actually in the Gorge itself.

A similar situation was observed among the small mammals. Typical montane species, (eg. *Lophuromys flavopunctatus*) were abundant on the plateau but absent from the Gorge. On the other hand, the Gorge mammal fauna included the spiny mouse *(Acomys cahirinus)* and gerbils *(Tatera valida)* typical of hot, dry regions and absent from the adjacent plateau. The striped mouse *(Lemniscomys striatus)* and the pigmy mice *(Mus proconodon* and *M. tenellus)* are typical East African savanna species present in the Gorge, but not on the plateau. The cane rat *(Thryonomys)* found at Forward Base 2 was the first record for Ethiopia, and is a species common in areas West of Ethiopia.

The small mammal species which formed permanent populations in the Gorge were noticeably thin on the ground. Only 52 specimens were caught in over 900 trap-nights; equivalent to 17 trap-nights per specimen (cf. 5 per specimen using the same methods and personnel in dry country

239

Chapman, C. M. 1970. Survey of the crocodile *(Cr. niloticus)* population of the Blue Nile. *Ibid.* p. 55–59.

Corbet, G. B. & Yalden, D. W. 1972. Recent records of Mammals (other than Bats) from Ethiopia. Bull. Brit. Mus. Nat. Hist. Zoology 22, no. 8: 213–252.

Hill, J. E. & Morris, P. 1971. Bats from Ethiopia collect by the Gt. Abbai Expedition in 1968. Bull. Brit. Mus. Nat. Hist. Zoology 21, no. 2: 27–49, 3 plates.

Johnson, E. D. 1972. Observations on the birds of the Upper Blue Nile basin. Bull. Brit. Ornith. Club 92: 42–49.

Snailham, R. 1970. The Blue Nile revealed. Chatto & Nindus, London.

16. THE BLUE NILE IN THE PLAINS

by

D. HAMMERTON

This somewhat longer stretch of the river from the downstream end of the gorge to the confluence with the White Nile at Khartoum contrasts strikingly with the previous sectors. The river enters a different landscape and a hotter climatic zone. From Lake Tana to the Sudan border at Deim the river has dropped over 1300 metres in about 860 km., while from here to Khartoum it only drops 130 metres in 740 km. There is still a turbulent and rocky upper part from Deim to Roseires including the extraordinary Fazughli Gorge where the river has cut a channel into the bedrock over 40 metres deep in low flow periods. By Roseires the slope has nearly levelled out and the river begins to meander through increasingly dry land.

One of the most striking phenomena of the Blue Nile is the change of colour at Khartoum from a deep green during most of the year to a reddish brown during the annual flood, a phenomenon well known to the ancient Egyptians, while even today the local population talks of the 'red water' arriving with great regularity. The name 'Blue' Nile is misleading and not a correct interpretation, the arabic 'Nil Azraq' meaning 'dark' Nile rather than 'blue'. The difference in colour of the two Niles at Khartoum is well shown in the colour plate (see Fig. 68).

Early travellers and explorers have left numerous descriptions, those of Bruce who came here in 1771 being perhaps the most impressive (see ch. 4c.). By contrast, only a few hydrobiologists (Ekman, Daday, Gurney, Chappuis) came in the early years of this century and the brief notes on the few water samples they could collect are now outdated and of historical value only.

The blue-green alga, *Anabaena flos-aquae f. spiroides*, was found, when in bloom, to be the origin of the green colour of the water – at that time an unusual discovery for river plankton (Rzóska *et al.* 1955). Obviously, the origin of this 'green water', passing by continuously, had to be sought upstream. A longitudinal study of the Blue Nile plankton was carried out between 1954–1956 with parallel examination of chemical and physical factors; to this were added some later observations by Talling in Ethiopia and the published results (Talling & Rzóska 1967) are summarized below.

Limnology of the Blue Nile before the Roseires Dam

Moving downstream of Roseires, not yet with a dam, a modest appearance

the water surface is 480 m above sea level and the capacity $3 \times 10^9 m^3$ but there is a design provision for ultimately heightening the dam to 490 m which would more than double the storage capacity to $7 \times 10^9 m^3$ (in contrast with the new capacity of the Sennar Dam of only $0.9 \times 10^9 m^3$). However, evaporation losses are considerable in this part of the Nile Valley and there is some controversy concerning the economics of heightening the dam.

The spillway and deep sluices of the dam have been designed to cope with the formidable problem of the Blue Nile flood with a maximum recorded flow of 10000 m³/sec. Impounding normally starts about October 1st and filling takes 3–4 weeks; drawdown commences in late December and the reservoir reaches its minimum level of 467 m by about the middle of May, see operational diagram (Fig. 56). This level must be maintained for the operation of the hydro-electric power station with a maximum generating capacity of 210 megawatts. The area of irrigation which can be served by the combined capacity of the Roseires and Sennar reservoirs, based on the criterion of an 80% flow year, is calculated at 3,512,000 feddans about 1400 thousand hectares), an increase of 1.3 million feddans over the area previously irrigated by the Sennar Dam. Stage II heightening to 490 m would increase the irrigable area twice, but would require a water allocation greater than available to the Sudan under the 1959 Nile waters agreement unless further capital

Fig. 54. The Roseires Dam in December 1966, shortly after completion; Photo D. Hammerton.

246

Fig. 55. The Roseires dam basin from space; Note the side canyons; ERTS.

works in the south are carried out to increase the natural discharge of the White Nile. Without such works the maximum irrigable area would be limited to a total area of 4.3 million feddans.

With construction of the dam under way in the early sixties, investigations were intensified from 1963 onward by D. Hammerton and collaborators at the University of Khartoum's Hydrobiological Research Unit, thus providing for the first time in the Nile system a study of biological changes resulting from the impact of a major dam on 640 km of river before and after its construction. Most of this work is so far unpublished, though salient features have been recorded in the Annual Reports of the Hydrobiological Research Unit, Khartoum; Hammerton (1972, 1973) has reviewed some general problems. What follows is a summary of the more important findings in contrast with conditions during 1963–66 (and 1951–56 discussed earlier).

succeeded in turn by *Anabaena flos-aquae* var. *intermedia f. spiroides* which was always dominant over other blue-green algae. In the 1966–67 season, *Microcystis flos-aquae*, together with the epiplanktonic *Phormidium mucicola*, became co-dominant with *Melosira granulata* in the Roseires basin. Previously this *Microcystis* has always been a minor constituent in the Nile. In the following season it appeared in both the Roseires and Sennar basins, while in 1968/69 *Microcystis flos-aquae* was the dominant element in succession to *Melosira* in the whole Blue Nile from Roseires to Khartoum. It is worth noting that this *Microcystis* was recorded during the 1966/67 season at the southern end of the Aswan High Dam basin at Wadi Halfa (Lake Nubia), 2000 km. downstream of Roseires, yet it was virtually undetectable between the two sites. In the 1969/70 season blue-green algae were almost absent at Roseires, whereas Chlorophyceae such as *Volvox* and *Pediastrum* appeared in larger numbers than recorded previously.

Hammerton (1972, 1973) drew the following conclusions from these investigations:

1. The Roseires reservoir has brought about greatly increased productivity in the Blue Nile by increasing the productive length of the river to over 700 km and by increasing the overall productivity as shown by standing crop measurements and primary production estimates.

2. The main development of plankton now takes place at Roseires where phytoplankton densities increase 10 to 200 fold within the basin. Although the density per unit volume was lower at Roseires than at Sennar, the density per unit of surface area was greater at Roseires because of the deeper euphotic zone.

3. It is suggested that the large increase in productivity of phyto- and zooplankton in the Blue Nile as a result of the two dams is analogous to increased productivity associated with eutrophication in many rivers in highly developed countries. At present the Nile is almost totally devoid of pollution and of nutrient enrichment from agricultural run-off. Hammerton believes that even a very mild degree of eutrophication from industrial development could have a serious effect on the Nile because of the high temperatures and high radiation inputs.

Zooplankton

Asim Ibrahim el Moghrabi (1972) studied the development of zooplankton along the river after 1966. Samples were collected at eleven stations along the entire course of 760 km of the Blue Nile from the entry of the river into the Sudan plains down to Khartoum. Sampling went on for four 'seasons', November until May–June when interrupted by the flood.

The results can be briefly summarized thus: 1. Upstream from the new dam, at 120 km distance, normal river conditions existed with a large

254

detritus suspension and few organisms, of which most were adventitious forms. At 60 km above the reservoir the suspension began to clear and true plankton forms were more evident. 2. Within the new Roseires basin there was a rapid development of zooplankton; with fluctuations; this state persisted in later years. 3. Downstream under the new hydrological regime, zooplankton densities were on the whole higher than recorded in 1951–56; considerable fluctuations occur. 4. Seasonal increases below the Roseires dam occurred after 1966 in February until May and in November/December at Sennar. At Khartoum the peaks of seasonal densities correspond largely to those observed in 1951–56, with time lags.

The composition of the Crustacean zooplankton (see ch. 24) is the same as reported a decade before by Talling & Rzóska. Fluctuations in dominant species have occurred but most of these are short lived. Moghrabi has investigated the species composition of the Rotifera and has recorded the presence of 26 species. Although it appears that reservoirs in tropical conditions reach some equilibrium much earlier than in temperate climates, yet it remains to be seen how future biological conditions will develop in view of the flood character of this river. A further study on zooplankton near the junction of the two Niles at Khartoum (Abu Gideiri 1969) alters only some details of species composition in the Rotifera. Seasonal changes and the two peaks of quantitative development found in 1951–53 (Talling & Rzóska 1967) are similar.

Moghrabi's thesis contains many graphs, both of general character and of particular species, forming a solid basis for comparing with any future development.

The Benthos of this river has not so far been investigated methodically. Our only, but very valuable, information comes from the work by J. Lewis and others, and deals with waterborn insects affecting human ecology (see ch. 23 and 25).

Hammerton (Ann. Report H.R.U. XIII: 8–9) reported the presence of large numbers of the giant bivalve mussel, *Etheria elliptica*, attached to rock outcrops in the bed of the river from Roseires to the frontier at Deim, a distance of over 100 km.

However, in May 1968 (Ann. Report H.R.U. XV: 8–9) he found that, as a result of the heavy deposition of silt within the dam basin, the entire population had been wiped out over a distance of several kilometres near the Fazughli Gorge. It seems that none of this species survives within the Roseires reservoir though live specimens have been seen attached to rocks downstream of the dam.

Fishes of the Blue Nile system

Except for some small speciation of *Barbus* in Lake Tana, the Blue Nile shows no differences in relation to the White Nile. From Khartoum to Roseires 46 species belonging to 14 families have been recorded (Abu

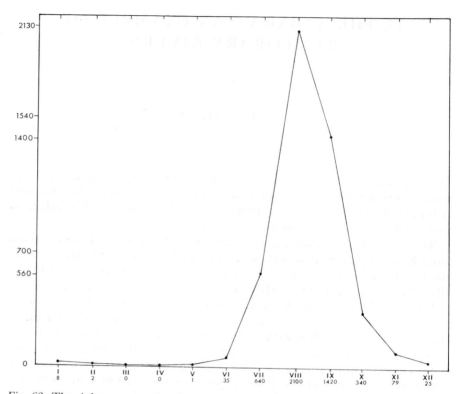

Fig. 62. The Atbara, example of a temporary river in the Sudan. Average monthly discharges 1912–1962 in m³/sec. The supply comes entirely from the Ethiopian mountains, nothing from its lower course in the Sudan; Note its dwindling to a trickle and complete dryness for 8 months. But its temporary contribution to the total water budget of the Nile system is considerable; based on figures from UNESCO/IHD 1969.

as biological habitat from descriptions by travellers and explorers in the past.

Bruce (1790) was there during his travels in Abyssinia and Nubia, 1768–1773. He was immensely struck by the combined effects of soil conditions and zonal-seasonal rainfall in the 'black earth' region in the south – the Sennar region – and the sharp contrast to these areas in the northern 'sands' of the Atbara lower stretches. During the '6 months of rain' in the south, conditions of life for the camel and cattle-owning 'Arab' tribes become impossible because of insect pests. He describes the 'tsaltsaya' or 'zimb' fly, probably either *Glossina* or more likely Tabainids (horse flies), which drive these tribes northwards towards the lower stretches of the Atbara which 'enjoy fair weather with no fly' and enough grazing for a few months. Even the wild fauna is involved; elephants and rhino of the upper and middle parts of the Atbara region are covered with 'tubercles' of insect bites. Migration is 'universal' and the insect pests are 'materially too interwoven in this country to be left as an episode'. The river itself in its middle stretches 'abounds' with hippopotamus and crocodiles. There were extensive woods around the river.

Baker, later Sir S. W. Baker, travelled a hundred years later in 1861–1862 and left a vivid description (1871, 1874). He arrived at the lower Atbara in May and found it drying out into pools. A veritable 'Exodus' of whole Arab tribes with thousands of camels and cattle was under way. Upstream, near Kassala, flooded clay plains are fertile but uninhabitable during the rains; the 'seroot' fly (Tabanids) was rampant and no domestic animal could survive in the Sofi region (near the Abyssinian border) except goats. Arabs have become nomads of necessity, caught between the need for pastures and the overwhelming flight from the 'fly'. The rainy season ends mid-September; after the grass has become dry in October, was burned and has begun to rejuvenate, people can migrate back to their homes. Baker describes giraffes on the upper Atbara 'teased' by 'flies' (possibly 'Pangonia' from his description) with flocks of birds flying around the giraffes and catching the insects. The big fauna migrates from south to north to confluence of upper Atbara and Setit.

Almost 80 years after Baker, in February 1939, Hurst investigated the topography of the Atbara region pursuing his hydrological exploration (1950). The water producing area of the Atbara lies 'almost entirely' in Ethiopia from the two main components and their tributaries descending

Fig. 63. Dinder river in the dry season; pools only are left with a concentration of zooplankton and in some cases a concentration of cat-fishes; Photo J. Rzóska.

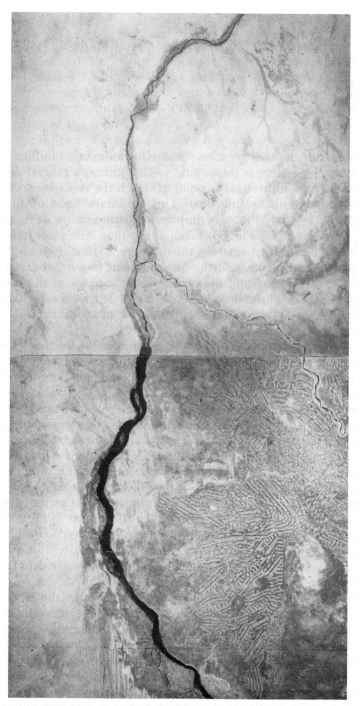

Fig. 64. Space-photograph of the junction of the White and Blue Niles. Also shown is part of the Gezira irrigation area with its network of canals, the Gebel Aulia dam basin when full, the twist of the main Nile at the Sabaloka gorge; ERTS.

Fig. 65. Airview of the river confluence at Khartoum; The Blue Nile is in the foreground, the White Nile in the background; Note the dark artificial 'oasis' of the Khartoum gardens (irrigated) against the bareness of the landscape; north is to the right; Photo Kuhn.

265

sp.? *Actinastrum schroeteri* G. Huber., *Dictyosphaerium pulchellum* H. C. Wood., *Gymnodinium* sp. possibly a form of *limneticum.*'

'The evidence points to the fact that most species of this Dongola plankton are derived from the White Nile, which as far as my own observations go, has its maximum phytoplankton population between November and December and this is usually dominated, as in the present sample, by *Melosira granulata* and the var. *angustissima*. There would, however, appear to be some Blue Nile elements, for from my comparative observations of the species present in the two tributaries, the following of the above list occured only in the Blue Nile: *Actinastrum schroeteri* and *Raphidiopsis mediterranea.*'

This has been quoted in full, because it is a proof that plankton blooms may occur in the Nile under natural conditions. At the time of taking the sample the regime of the previous Aswan dam extended its influence only to north of Wadi Halfa; Dongola is about 400 km or more upstream from this point.

Years later in 1964, D. Hammerton, leader of the Hydrobiological Research Unit collected data on the biology and chemistry of this northern river; unfortunately the biological samples were largely lost during a serious road accident. Some few plankton samples which survived gave a much poorer plankton picture than that obtained by Rzóska; Hammer-

Fig. 70. The drowning of the town of Wadi Halfa by the rising waters of the Aswan High dam basin; only the top of the minaret is still visible in 1964, soon to disappear completely; D. Hammerton.

268

ton ascribes this (Ann. Rep. Hydr. Res. Unit 1963/4 no. 11:8–11) to gale conditions with considerable quantities of sand blown into the river. In my opinion such changes must be expected; the samples were taken at a different time of the year (March/April) and secondly rivers show 'bouts' of biological conditions in a moving water mass.

Hammerton's observations on the phytoplankton and limnology will be discussed by Talling in ch. 26 and 27.

All these observations on the Nubian part of the Nile have now only historical value, besides being clear indicators of *potential* developments, to be expected under the new environment of the Aswan High Dam basin. This reaches deep into the Sudan for about 170 km and had drowned the town of Wadi Halfa (see Fig. 70), all riverain cultivations and the dwellings of the riverain people. With the gradual rise of level in the reservoir, which is of century storage, Lake Nubia will ultimately have an area of 1,400 km². Burried under its water are sites of immense prehistoric and historical value and prominent geomorphological features, like the famous II cataract.

All the events which can now be predicted confidently, have happened here: the enormous sedimentation caused by current slow-down, the clearance of water and subsequent development of primary production in the form of phytoplankton blooms. Chapter 19 deals with these developments in some detail for the whole dam basin. Here are some observations made in Sudanese Nubia from June 1967 to June 1968 on the incipient fisheries (T. George 1973). In the context of this book they are used as one of the biological indicators of the new environmental conditions.

During the two flood seasons observed water levels rose between the end of June to mid-July, but the brown water mass of the flood reached Wadi Halfa only by mid-October. The water became clear by the end of December and the phytoplankton began to build up with a peak in February–March. Temperatures of surface waters reached a maximum of 31° C in September and fell to its minimum of 8 °C in January. These extremes, imposed by the desert climate, contrast sharply with those of inland Africa.

According to the results of the fishing party led by T. George 36 species of 14 fish families were recorded; it is worth noting that in 1964 a Swedish expert (Mathiasson) found 27 species of which 10 were not seen in the 1967/8 catches. The publication by T. George deals with the analysis of the fish catches, in details of length/weight relations of particular species, catch and abundance in the various months and some remarks on the food of the 8 main species. As the lake is so young it has not yet developed biologically; at the time of investigation there was no shore flora nor was these any mentioning of a shore fauna and deeper benthos. The full adjustment of the nilotic fish fauna and all other components of such newly created lake will take some time. Meanwhile the 'preliminary account' of 1967/8 is of great value not only for practical reasons but also as record against which future changes may be measured.

REFERENCES

George, Th. T. 1973. Preliminary Account of the Fish and Fisheries of Lake Nubia. 1967–1968. J. Indian Fisher. Assoc. 1: 65–88, 1971. (Reprint 1973).

Hammerton, D. 1970. A longitudinal survey of the main Nile. Hydrobiol. Res. Unit Univers. Khartoum, 11 Ann. Rep. 1963/4: 8–11.

Mathiasson, S. 1964. (Quoted after George).

Rzóska, J. 1958. Notes on the biology of the Nile north of Khartoum. Hydrob. Res. Unit, Univ. Khartoum, Ann. Rep. 5 1957–58: 16–20.

19a. LAKE NASSER AND LAKE NUBIA[1]

by

B. ENTZ

Introduction

Lake Nasser forms together with Lake Nubia, the southern part of the huge reservoir, one of the greatest man-made lakes in Africa. This artificial impoundment on the Nile River when full, will extend from the High Dam (HD) at Aswan (Egypt) till the Dal Cataract (Sudan) between the 23°58′ and 20°27′ north latitude and 30°35′ and 33°15′ east longitude. Construction started in 1960, filling in 1964. (Map, fig. 71).

It is unique in its performance because it is situated in pure desert surroundings where the yearly mean precipitation does not surpass 4 mm/year, and a very high rate of evaporation around 3,000 mm/year. The air is very dry and the sky is almost completely cloudless. The only source of water is the Nile River with its inflow in the south. The outflow at Aswan is the continuation of the Nile River towards the north. This vast impoundment is in reality not a typical lake but rather an extremely slow flowing river. Some of the most important morphometric data of the reservoir are summarized in table 1.

As obvious from this table and from Fig. 72 and 73, the lake has a long, narrow shape with often dendritic side areas, called 'khors'. The number of important khors is one hundred, 58 of them being located on the eastern and only 42 on the western shore. The total length of the khor systems in the lake, when full, will be nearly 3,000 km and their perimeter will surpass 6,000 km (Fig. 74). The total surface area of the khors that is areas outside the main valley covered by water, is about 4,900 km² = 79% of the total lake surface but in volume they will contain only 86.4 km³ water (= 55% of the total lake volume).

The mean slope of Lake Nasser is steeper on the generally rocky or stony mountainous eastern shore (14°07′) than on the flatter, more open, wider, often sandy western one (1°51′).

[1] All original data pertaining to Lake Nasser and Lake Nubia were obtained by me as Project Manager – Fishery Limnologist at the FAO/UNDP/SF Lake Nasser Development Centre Project. I am indebted to express my deepest gratitude to the Government of Egypt, especially to the Regional Planning of Aswan and personally to H. E. M. Alwany Governor of Aswan, Mr M. Marei General Director of RPA and Dr. A. F. A. Latif Project Co-Manager, Director of the Lake Nasser Centre of RPA and last but not least to all my colleagues and friends whoser eadiness and permanent help made this work possible.

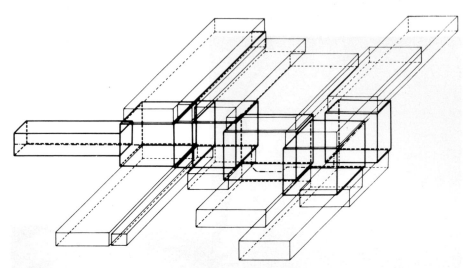

Fig. 73. Schematic representation of the Aswan High dam basin. –.–.– Main axis of the reservoir; ⎯⎯ volumes of central parts of sections 8-1, sections 1 and 2 are combined (thick lines) ; ⎯⎯ volumes of side areas and khors (thin lines) ; Entz unpublished.

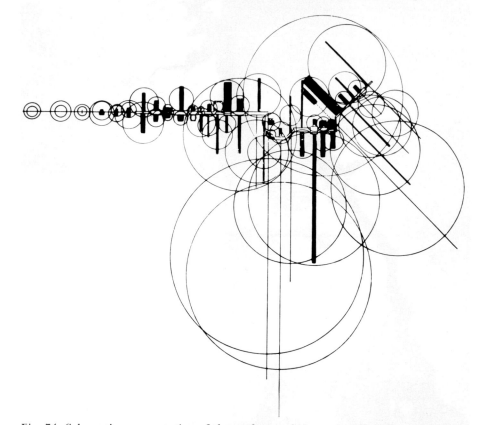

Fig. 74. Schematic representation of the surface and the perimeter of the subsections of the lake. ☐ surface of central areas; ■ surface of side areas and khors; ○ perimetersof the side areas and khors. Entz unpublished.

Table 1. Morphometric data of Lake Nasser and the total reservoir.

Water level	Lake Nasser		Total reservoir	
	160 m	180 m	160 m	180 m
Length km	291.8	291.8	430.0	495.8
Surface area km²	2585	5248	3057	6216
Volume km³	55.6	132.5	65.9	156.9
Shoreline km	5380	7844	6027	9250
Mean width km	8.9	18.0	7.1	12.5
Mean depth m	21.5	25.2	21.6	25.2
Maximal depth m	110	130	110	130

The mean length of the main khor systems is increasing from the south ($\pm$ 1 km between 500 and 360 km from the HD) to the north ($\pm$ 43 km between 80 and 0 km from the HD) because of the northwards declining ancient river bed.

Due to the dendritic form of the lake basin the shoreline development value (D_L according to Hutchinson) is extremely high (33.1!) as compared with usual values of lakes usually ranging between 1.8 and 6.0.

All the khors have a 'U' shaped form in their cross section with a flat sandy central belt and sometimes, especially on the eastern shore, very steep (slope up to 48°) and mostly rocky shores.

Morphology and hydrology

The deepest part of the lake is the ancient river bed with the adjacent strips of cultivated land forming together the original river valley, called the central area of the lake with its bottom elevation between 85 and 150 m a.s.l. The side areas lie between 150 and 180 m above sea level.

The central part can be considered as a flowing river-lake where the speed of the current is fast at the southern end of the Nubian gorge region (100 to 150 cm/sec). This speed is gradually reduced within a few kms to 10–20 cm/sec and in Lake Nasser to 0–3 cm/sec. The mean depth of this central part is gradually increasing from 10 m at the southern end to 70 m in the north. The bulk of the water masses coming from the south is passing through this central part, which forms about half of the total volume of the lake.

The flood, which arrived at Aswan from Khartoum within one month before the High Dam was built, covers now the same distance in not less then 5 but sometimes probably more than 12 months, depending on the lake level and the strength of the flood.

The side arms show usually lacustrine characteristics. Currents if present are weak and are most probably induced mainly by the wind and

Table 2. Characteristics of Lake Nasser and L. Nubia in the different subbasins.

		Subbasins		
	A	B	C	D
Area	Dal-Attiri	Attiri-II Cataract	II. Cataract-El Dirr	El Dirr-High Dam
Distance f. HD km	500–410	410–365	365–200	200–0
Per cent of length of reservoir[1]	18	12	30	40
Per cent of surface of reservoir[1]	1.70	1.34	33.00	63.96
Per cent of volume of reservoir[1]	0.24	0.63	33.77	65.36
Main character during flood	Riverine[2]	Riverine	Semi riverine[3]	Lacustrine[4]
Main character, during low water	Riverine	Semi riverine	Lacustrine	Lacustrine
Approximate speed and main type of current in cm/sec during flood	200–120 turbulent	110–80 turbulent	12–8 laminar	3–0 laminar
Approximate speed and main type of current in cm/sec during low water	100–60 turbulent	50–20 laminar?	5–3 laminar	2–0 laminar

1. All values calculated for 180 m lake level above sea level.
2. High inorganic turbidity, turbulent currents, no stratification.
3. Both inorganic and organic turbidity, stratification weak or of short duration.
4. Dominant organic turbidity, weak laminar currents, long lasting stratification from April to October or December and total circulation from November or January to April.

Table 3.[1] Lake levels between 1964 and 1974, as in a.s.l.

	1964	1965	1966	1967	1968	1969	1970	1971	1972	1973	1974
Maximum (December)	127.8	132.5	142.5	151.1	156.4	161.7	164.5	167.2	165.3	166.7	170.0
Minimum (July)	106.2	116.0	119.3	130.2	145.5	150.7	153.8	159.8	162.6	158.2	161.0

1. Data from the High Dam Authority, Aswan.

far less by the progressing water masses of the central part of the reservoir. According to the current conditions the total reservoir can be divided into four subbasins. Along these, the original riverine conditions are gradually lost and the reservoir resembles more and more a typical lake. Some of these features are summarized in table 2.

The lake's expected high level of 180 m a.s.l. has not yet been reached. The lake was half full in 1969 at about 160 m and the filling process continued up to 1971. At that time influenced by the extremely dry weather along the Blue Nile the filling process was interrupted and the water level dropped in the following period, but the increase was regained again in 1973–1974 (see table 3).

As a consequence of the yearly flood, the maximum lake levels can be measured in December or January and the minimum values in July, just before the start of the new flood period.

With the huge buffer capacity of the lake, the course of the Nile flow below Aswan is basically changed. The one-peak flood of different amplitudes between September and January and the minimum water level in July have been replaced by an almost constant artificial medium water supply between May and June, adequate for irrigation demands for summer crops and by a second lower peak in December for the winter crops besides a fairly constant water level in the remaining periods.

Consequently the amount of water actually entering the Mediterranean is reduced considerably though not completely but the sedimentation of silt load in the delta area is practically abolished.

The three main factors affecting the conditions of the reservoir are air temperature, the yearly flood of the Nile and wind conditions, causing together with biotic factors basic changes in water temperature, oxygen saturation, optical conditions, conductivity, changes in dissolved chemical components and even changes in the fish yield.

THE NILE FLOOD

Sedimentation of Nile-mud within Lake Nasser and Lake Nubia.

As widely known, the Nile carries a heavy load of mud of about 100 million m³ = 0.1 km³ per year, consisting of a mixture of sand, silt and clay. This mud was an important source of fertility for Egypt for thousands of years.

During the first years the suspended mud was 'flushed' through the basin of Lake Nasser. But after reaching 150–155 m lake level in 1967–1968, the speed of the current within the lake was reduced so much that the previously dominant turbulent currents were replaced by laminar ones, and a strong sedimentation started within the lake basin. As a visible sign of this process since 1969, the Nile water became permanently green and transparent around Aswan, because of the almost complete lack of suspended silt.

278

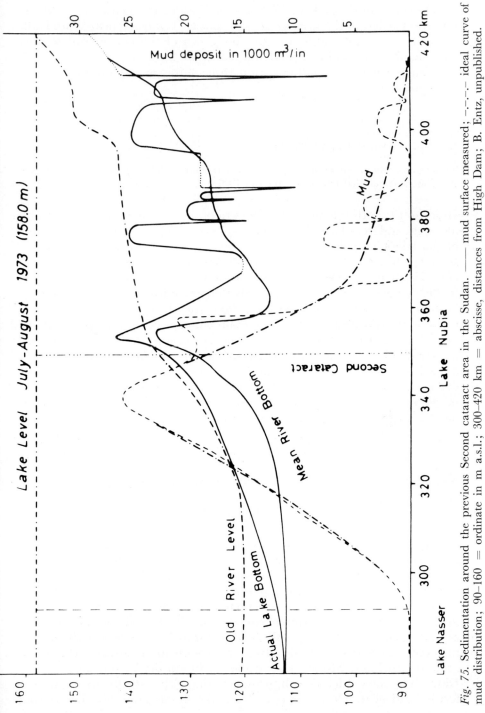

Fig. 75. Sedimentation around the previous Second cataract area in the Sudan. —— mud surface measured; –·–·– ideal curve of mud distribution; 90–160 = ordinate in m a.s.l.; 300–420 km = abscisse, distances from High Dam; B. Entz, unpublished.

Following the special shape of the sub-basins, the center of sedimentation is located in the area of the previous II. Cataract, near Wadi Halfa where a very remarkable abrupt slow down of the current takes place with the formation of a new 'delta' (see table 2).

From the point of sedimentation three subsequent sections could be distinguished:

1. Between Attiri and Gomai   (410–365 km from HD)   31.77%[1]
2. Around the 2nd Cataract   (365–345 km from HD)   20.06%
3. Between Wadi Halfa and Abu Simbel   (345–285 km from HD)   48.17%

[1] Foot note: The percentage values given correspond to the proportion of the amount of sedimented material between 1967 and 1974.

In section 1. both thickness and amount of the sediment are very variable along the different localities depending on the width and depth conditions of the lake channel. There are locations where remarkable amounts of sediment have been accumulated but also others without new sediment (Fig. 75). The peak was reached in section 2., where a thick sediment (up to 20–25 meters) covers almost evenly the previously very uneven cataract bottom. In section 3. there is a very smooth decrease both in thickness and the amount of sediment along the main channel beginning with a thickness of about 10 m, to almost nil. The maximum amount of suspended matter within the years 1968–1973 was ranging during the flood between a few hundred mg/l and 8,000 mg/l of perhaps more. The extreme value of 8 g/l has been measured in 1973. This high value never recorded before might be in connection with the exceptional flood of 1972/73 and the newly constructed reservoirs on the Blue Nile and the Atbara.

Future aspects

Naturally new amounts of suspended matter of an approximate amount of 0.1 km³/year – will be sedimented in the lake. Because of this the lake will undergo in the neighbourhood of Wadi Halfa obviously basic changes in the near future. As suggested the lake will be gradually shallower in this area until it reaches a depth of about 6 to 10 m. This can be promoted or delayed by the tempo of the filling process of the lake and the strength of the flood, until the suggested 'working level' of the lake will be reached. If the supposed 6–10 m river depth will be attained, possibly in 5 to 15 years, the start of a new river channel formation can be expected in this area. Beside the channel the appearance of shallow mud banks is probable, emerging from the lake surface during low water periods. Similar phenomena could be already observed in some areas towards the south, around Attiri and Murshed at 415 and 370 km distance from the High Dam. The new 'delta' will grow continuously northwards and the fertile mud of the Nile will cover the lake shores gradually like during the last years between Attiri and Wadi Halfa, where newly deposited mud appears at low water level on the shores. This new layer is in some calm beaches with flat slopes 1 to 3 m thick, covering step by step the previous riverside vegetation. Its thickness is decreasing inland and is nil at the shoreline of the highest lake level.

This drying mud layer is often covered by sand blown by the wind. The total amount of deposited mud is diminishing towards the north and its last appearance is for the time being around Abu Simbel in the form of a thin mud film. The sedimentation of mud will continue gradually towards the High Dam, but it will take most probably several hundred years until the deposits will approach the Aswan area in remarkable amounts.

TEMPERATURE

The tropic of Cancer crosses the lake; the major part is situated within the tropics and only a small proportion in the northern mediterranean zone. Lake Nasser is a monomictic subtropical lake (Hutchinson 1957) with prevailant lacustrine characteristics and with a single circulation

Table 4. Temperature values of Lake Nasser within the period 1970–1974.

	1970	1971	1972	1973	1974
Minimum water temperature measured °C (February-March)	17.6	16.3	16.1	15.7	15.2
Approximate mean temperature °C (February-March)	18.5	18.4	16.8	16.9	15.9
Lowest temperature °C measured in July-August	19.8	18.6	18.5	18.0	16.7
Approximate mean temperature °C in July-August	25.0	23.2	23.8	22.8	22.9
Volume of water with temperature below 20 °C 'cold water' in July-August km³	0.02	4.3	7.0	6.2	10.7
Per cent of volume of 'cold water' of total lake volume in July-August	0.1	7.7	12.2	12.3	18.4
Maximum temperature measured in the open water °C	32.2	29.7	34.5[1]	33.0[1]	31.8
Minimum depth under the lake level with temperatures below 20 °C in July-August, m	60	37	37	34	26
Summer stability between 0 and 40 m depth in mkg (mean values) per square meter	157	195	186	208	271
Approximate depth of metalimnion during stratification in summer m	16.0	13.4	18.0	11.1	13.1

[1] Foot note: Temperature values measured in Lake Nubia.

period in winter between November and March. Lake Nubia, and especially its gorge region (Sub-basins A and B) bear mainly semi-riverine or pure riverine features, where the temperature is greatly affected by the current of the inflowing Nile water. Some of the temperature and stability data for the years 1970–1974 are summarized in table 4.

As a result of the falling air temperatures in autumn, the total water mass is cooling down and a complete circulation develops in the winter months. It is worth mentioning that the mean temperatures calculated and the actual minima measured were gradually decreasing from 1970 to 1974. The absolute minimum (15.2 °C) could be observed on 18/2/1974 in deep water near the High Dam (see table 4). The same trend could be observed as decreasing minima in summer time and as an increase of the amount of the 'cold water' during the summer months (table 4). This cooling down of the lake might be continued in the future in some extent mainly due to the low mean ambient temperatures of some cold days in January ($\pm$ 14 °C), mostly accompanied by increased evaporation induced by sometimes long lasting cold mortherly winds, though lower water temperatures than 13 °C can hardly be expected. Both the amount and the proportion of 'cold water' (15–20 °C) in the deep water layers during the summer stagnation period can still increase significantly in particular under the influence of rising water level. The final picture can be clarified only after several years.

The actually measured maximum lake temperature was 34.5 °C. This could be observed in the open water of Lake Nubia under stratified conditions, just below the surface, at 3 p.m. during a long lasting windless period, which permitted the development of a very heavy Cyanophyta water bloom. The maximum water temperature measured in 1 m depth was still 31 °C, but that in 2 m depth only 29.5 °C.

Accordingly, the range of water temperature does not exceed 20 °C, which is less than the usual range of 22 to 30 °C in shallow temperate lakes and of course much less than the extreme values of the ambient air temperature (—5 to +52 °C). These low surface values are surprising in the 'land of eternal sunshine', and can be explained only by the very high evaporation rates. There are big differences in the circadian air temperature up to 25 or 35 °C a day. But this affects only the temperature of the upper water layers.

The course of the temperature of the whole water mass is mainly influenced by the seasons. There is a rapid increase of the air temperature from April to June followed by that of the water. Near the southern end of the lake under riverine conditions the water is gradually warmed up in all depths to 28–30 °C. A quite different feature could be observed in the lacustrine sub-basins of the lake between the Second Cataract and the High Dam. Here only a slight gradual temperature increment of short duration could be noticed all over the total water mass. After that – usually already in April – a short but rapid increase takes place in the upper 10–15 m water layer resulting in the summer stratification. Although the depth of the metalimnion is changing during the summer seasons and is different in the various localities, it laid usually between 10 and 16 m. Under the influence of the dominant winds and the inflowing river, the depth of the epilimnion is not uniform in the whole lake. As a good example the conditions at the end of September 1970 could be mentioned when the metalimnion was at 15 m in Khor El Birba (9 km from the HD), at 25 m in Allaqi (100 km f. HD), at 30 m around Singari (in the valley bend 180 km f. HD) and at 50 m at Adindan (300 km f. HD).

The slope of the temperature boundary can be well compared with the stability values expressed in mkg/m² (Ruttner 1933, 1959). These values reach between surface and metalimnion only about 10 mkg/m². The same values are high between surface and 40 m depth as compared with deep temperate lakes (e.g. Altausseer-see 60 mkg/m²) or even tropical lakes (e.g. Ranu Klindungan in Sumatra 190 mkg/m²), ranging from 157 to 271 mkg/m², but showing extreme values up to 325 mkg/m². Between surface and 55 m depth very high stability values could be calculated up to 700 mkg/m² or even more. High stability is attained very fast, already in June and July, but from the second half of August it is again decreasing due to the diminishing ambient air temperature and the turbulent water movements caused by the approaching flood. As a result the location of the thermocline drops down in time, especially in the south and disappears completely in October or latest in November. Accordingly the stability values are decreasing until they are reduced to zero under winter conditions (isothermic conditions).

DISSOLVED OXYGEN

The oxygen saturation reflects the influence of four important factors: the temperature regime, the wind conditions, aquatic life activities and the desert surroundings.

In winter time there is a total circulation present in the lake. The water is oxygenated from the surface to the bottom and the saturation ranges usually between 60 and 120% as in February (see Fig. 76). The conditions observed during this period were very similar from the neighbourhood of the High Dam through the central section of the lake to the southern gorge region.

Parallel to the rapid increase of the water temperature in April a remarkable increment of the O_2 saturation appears in the upper water layers. This phenomenon is most pronounced within the northern areas of the lake in 1 to 10 m depths where saturation values above 150 or even 200% have been measured. This could happen in a windless period under the influence of a very rich phytoplankton, mostly Diatoms. A similar trend of O_2 saturation could be observed in the central

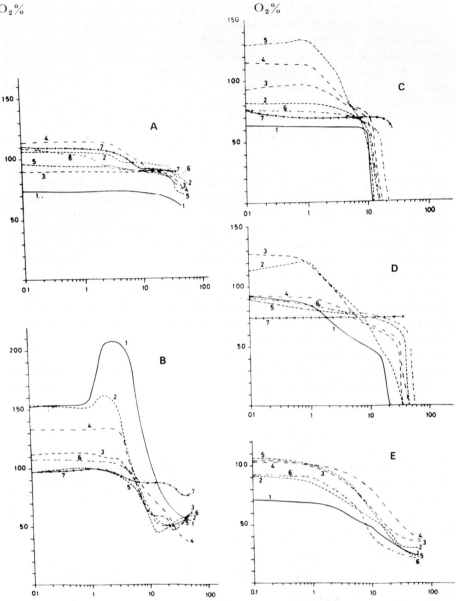

Fig. 76. Oxygen saturation v. depth in m in different seasons and localities of the lake; A. February; B. April; C. July; D. September; E. November; 1. High Dam Area, 0–7 km; 2. El Birba-Allaqi, 9–100 km; 3. Sayala-El Sibu 110–160 km; 4. Shaturma-Kurusku 165–190 km; 5. El Dirr Tushka 200–240 km; 6. Adindan-Wadi Halfa 270–350 km; 7. Amka-Murshed 355–390 km; distances from dam; B. Entz, unpublished.

parts of the lake, although in a smaller estent. Some higher saturation values present in Lake Nubia were caused by heavy Cyanophyta water blooms. During this period the total circulation was already blocked and the beginning of the formation of the metalimnion was obvious. The oxygen saturation drops rapidly from about 5 to 12 or 25 m depth, but grows again in the deeper layers (Fig. 76). The probable explanation of this peculiar slope of these curves is suggested as follows. The oxygen saturation of the lower layers starts to decrease as soon as the total circulation is interrupted. This process is the most vivid just below the zone of the metalimnion under formation, because of the high O_2 consumption by plankton and nekton organisms moving upwards and the rapid decomposition of plankton rain sinking down into the hypolimnion. The bottom of the lake does not consume so much oxygen as usual in lakes because it is mainly formed by desert soil without any remarkable decomposing organic admixtures. The oxygen saturation does not fall anywhere below 50 or 60% saturation in this period.

Quite different are the conditions of the southernmost 'riverine section' of the lake, where all water layers are mixcd throughout the year and the oxygen is almost evenly distributed.

During the following weeks a stable stratification is formed by further rise of the temperature. The result is a well oxygenated epilimnion and an oxygen free hypolimnion. During the summer months a remarkable increase in the oxygen saturation takes place in the epilimnion towards the south, most probably due to the masses of Cyanophyta, causing severe water bloom especially in Lake Nubia.

The stratification is abolished gradually in the autumn all over the lake progressing towards the north with declining ambient temperature caused by the winds and the inflowing cooler flood water in the south. A complete total circulation is again reestablished with gradually increasing saturation values during December, January or February.

It is worth mentioning that the oxygen conditions showed no remarkable annual changes during the period 1970–1974. Thus the seasonal results of different years are comparable.

OPTICAL CONDITIONS

The transparency of the lake is affected by three important factors: 1. the inflowing turbid water of the Nile River; 2. the development of phytoplankton and; 3. vertical water movements (wind action).

The inflowing Nile water but especially the flood water is very turbid and has a brownish-greyish colour. On the arrival of the flood into Lake Nubia the Secchi transparency can be diminished within a few hours from 70–140 cm to 20–30 cm oɪ, as in the southern part of it, even to 5 or 10 cm. The border line between flood water and old water is sometimes very sharp as north of the Second Cataract area where the current is

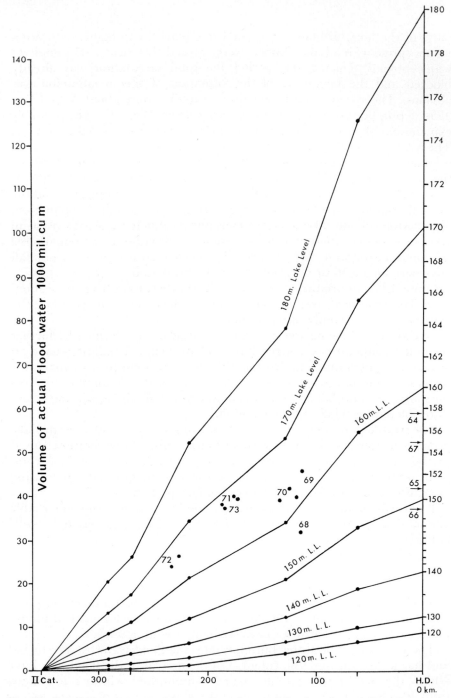

Fig. 77. Correlation between lake levels, volume of flood and the edge of the area reached at New Year by the turbid water; ● actually observed; ○ calculated site; B. Entz, unpublished.

already slow (5–10 cm/sec), but the silt content is still remarkable (100–500 mg/l).

Along the progressing flood a continuous sedimentation takes place within the lake, accompanied by gradually reduced turbidity. Ultimately the optical border line between flood water and old water disappears. By this way the progress of the flood can be followed within the lake by Secchi-disc measurements, but only to a certain distance depending on the hydrological conditions.

The time of flood-water penetration within the lake depends on two factors: 1. the amount of inflowing flood water from the beginning of the flood until the date of observation (straight correlation), and 2. the actual lake level at the beginning of the flood (inverse correlation). If these two components together with detailed morphometry of the lake are known the way of the flood water can be followed in the lake by calculation. Even the actual border line between turbid and clear water at New Year can be calculated by multiplying the volume in question with the empirical factor 0.64. This factor is a result of sedimentation of silt particles during the period considered.

The area of turbid water at this time is of practical importance for this area and is in straight correlation with the distribution and catchability of some semiriverine fish (e.g. *Alestes* spp. and others) in the lake. Based on the above mentioned data the probable amount of catch of the species mentioned can be predicted every year for the next season. The empirical and calculated locations for the border line of the flood water and old water are in good correlation and their position at the New Year are presented for the years 1964 to 1973 in Fig. 77.

In areas where the water masses contain only low amounts of silt i.e. in which the sedimentation has already been completed there is a permanent high transparency in the deeper water layers. It can be suggested that there the Secchi values would exceed all year round 300 to 600 cm or even more. In these areas the transparency of the epilimnion is controlled mainly by the phytoplankton. If the phytoplankton density is poor, as usual from December to February, the transparency ranges between 200 and 400 cm. As soon as a remarkable algal development starts, usually in March or April, the transparency will be reduced to 80–130 cm or as in case of a dense water bloom even to values of 50 to 70 cm.

Under special conditions i.e. strong, long lasting wind blowing from the shore towards the deep water areas of temperature homogeneity, an upwelling current can be formed easily, lifting water masses very poor in plankton to the surface (see Fig. 78). Under such conditions on 18/3/1974 an extremely high transparency of 745 cm could be determined. After a few days of calm weather the transparency dropped because of renewed and rapid phytoplankton development and the conditions became normal again.

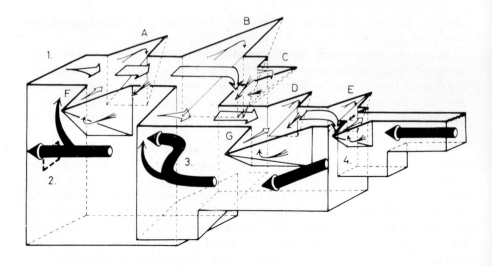

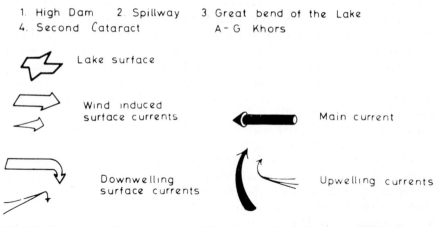

1. High Dam   2. Spillway   3. Great bend of the Lake
4. Second Cataract   A – G Khors

Lake surface

Wind induced
surface currents

Main current

Downwelling
surface currents

Upwelling currents

Fig. 78. Current conditions in Aswan High dam basin; B. Entz, unpublished.

In the absence of large amounts of suspended silt, the lake has a green or yellowish green colour. Under such conditions 1% of the surface light reaches only 2–6 m depth. These layers are usually deeper in winter and early spring and shallower in the other seasons. In the brownish or greyish flood water the same light proportion does not reach more than 100 or often only 40 or 50 cm depth.

THE WIND

One of the major factors affecting the lake conditions is the wind. Based on regular meteorological observations in Aswan and in Wadi Halfa (Omar & El-Bakry 1970) and on our own studies during 1969–1974, the present knowledge on the wind conditions can be summarized as follows. The dominant wind direction is from N–NW (83.3% of windy days), blowing along the main channel or towards the lake centre in the khors of the western side and towards the shore in the opposite eastern khors, causing leeward surface currents in the different areas.

The speed of strong winds reached usually 10–15 m/sec or sometimes 18 to 20 m/sec. Heavy storms well known in several tropical regions have not been noticed in the area during our work. There were stronger winds in wintertime, causing complete circulation. In summer months only moderate winds were common, interrupted frequently by shorter of longer calm periods especially in July or August which promotes a stable stratification. As an average in July 46.4% of the days are calm (Aswan Meteorological Reports 1926–1955). In autumn under cooling weather conditions the wind is helping to destroy the lake stratification.

Moderate but sometimes remarkable wind induced surface currents are detectable in the lake down to 3 or 5 m depth. Their direction in the central channel is from north to south i.e. just opposite the main south-north current of the lake, with a massive flow of water following the Nile valley towards the High Dam (see Fig. 78).

In special localities of the reservoir the existence of wind induced upwelling water movements could be suggested. These were replacing the surface water masses blown away by the wind pressure. Areas with upwelling currents could be characterized by reduced water temperature and oxygen saturation, lower pH values and high, sometimes surprisingly high transparency values of up to 600 or probably 800 cm Secchi values.

Such conditions could be found frequently near the High Dam and sometimes at the southern end of the valley bend near Amada (200 km f. HD), and almost regularly at the end of khors on the western shores. Wind induced sinking water movements were present in other localities resulting in the accumulation of warmer water masses rich in oxygen with high algal turbidity and high primary production. This phenomenon could be observed in erosion littorals e.g. in khors of the eastern shore and in particular in the main channel at the northern end of the bend of Lake Nasser (Singari & Kurusku, 180 km f. HD) as also in the gorge region of Lake Nubia, near the previous Second Cataract (360 km f. HD). Such peculiar conditions diminished the depth of the metalimnion under upwelling and increased it under descending water movement.

The speed of wind induced currents could reach 10 to 35 cm/sec. The above scheduled system of currents covering practically the whole surface of the reservoir (Fig. 78) may be an effective way to avoid any gradual increase of salinity despite the extremely high rate of evaporation.

Another action of wind or even a slight breeze is its remarkable direct cooling effect on the surface water temperature. Wind is enhancing the already high rate of evaporation, under conditions of extremely low relative humidity (of $\pm$ 35% in winter and only 13–21% in summer), to about 3,000 mm/year. These phenomena seem to explain reasonably the relatively low surface water temperature (25–33 °C) under extremely high ambient temperature of the surrounding desert (40–52 °C) in summertime.

CONDUCTIVITY

It is well known (e.g. Hammerton 1969) that there is a remarkable fluctuation concerning the electric conductivity of the Nile waters. As soon as the flood water appears in Lake Nubia, the conductivity drops down in the area affected from about 280–300 μmhos to 220–230 μmhos cm followed by a more gradual decrease to 210–160 μmhos or even lower.

As mentioned above the total reservoir and particularly its main channel can be considered rather as a very slow flowing river, with predominant laminar currents, than as a typical lake. Accordingly the opportunity arises to study the progress of the flood by fast, simple and exact conductivity determinations, besides transparency observations. Some of our results obtained by conductometry, are shown in Figs 79a and 79b and are calculated as geometric means of the vertical measurements or of all the samples of the cross sections concerned. It is worth considering how the water masses of equal conductivity are moving evenly from September to November 1970 along the main channel, and how far the volume of the lake section concerned and those of the moving water masses are closely correlated. As evident from Fig. 79a, the water masses with 235 μmhos conductivity present at Adindan (280 km f. HD) in September (A) reached Madiq (135 km f. HD) in November (A'). During the same period waters with values of 265 μmhos (A$_1$) moved from Tushka (240 km f. HD) to Allaqi (110 km f. HD; A$_1$'), or those of 290 μmhos (A$_2$) from Singari (185 km f. HD) to Mirwaw (70 km f. HD; A$_2$'). It could be concluded that during the period in question it took about six months, for the front of the flood to pass through Lake Nasser from Adindan to the High Dam. This suggestion could be confirmed by several biological phenomena.

The data of the flood period 1973–1974 show very similar characteristies (Fig. 79b). During this period the movement of the flood water and as that of the low water can be followed from July 1973 to July 1974 similarly as explained above for 1970, though this time the process is not so clear. The extreme values became reduced in time, and the borders of the different types of waters are not so distinct, most probably due to the meanwhile increased water level.

Conductivity measurements give often useful detailed information concerning the water movement within any selected cross section, such as

290

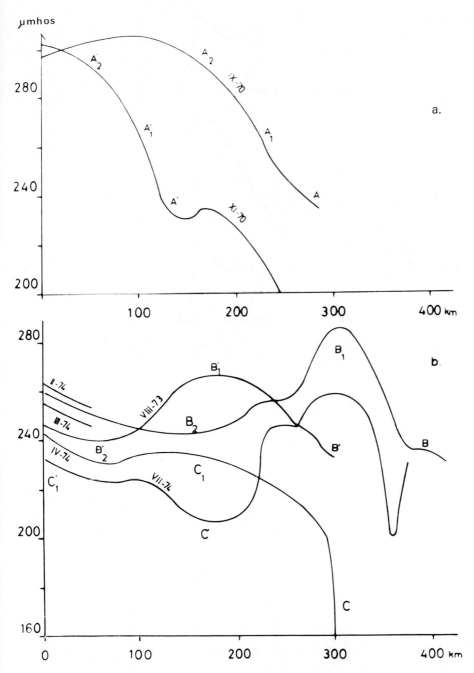

Fig. 79. Conductivity conditions as affected by the flood; a. in 1970; b. in 1973–1974; B. Entz, unpublished.

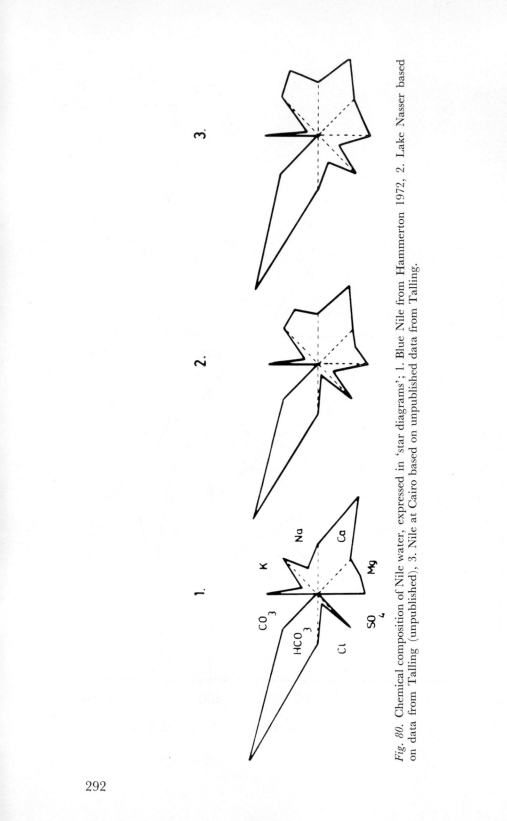

Fig. 80. Chemical composition of Nile water, expressed in 'star diagrams'; 1. Blue Nile from Hammerton 1972, 2. Lake Nasser based on data from Talling (unpublished), 3. Nile at Cairo based on unpublished data from Talling.

vertical and horizontal location of the main current, areas rich or poor in plankton and the water movements in the khors.

Water chemistry

The basic ionic composition of the lake water does not differ from the inflowing Nile water. This can be easily seen in Maucha's star diagrams in Fig. 80 (Maucha, 1948). Within the lake only a slight increase occurs in the Na^+ and Cl^- contents and an almost negligible decrease of the Ca^{++} content from the inflow to the outflow.

The range of pH is wider than in the main course of the river, and due to intensive assimilation processes in the surface layers pH values up to 9.6 are present. According to the studies of Nessiem the phosphate contents together with nitrate are decreasing from the south but a slight increase in nitrite was found (Nessiem 1973).

During the summer stagnation period some H_2S appears in the hypolimnion, detected just below the metalimnion probably due to the decomposing plankton rain, though the S^{--} concentration even here remains below a few tenths of milligrams per litre. The surface of the bottom within the old river channel is brownish in the south, but becomes black in the north during the summer stagnation period, probably because of FeS accumulation. Interesting changes of suspended and dissolved organic matter appear in the Second Cataract area (Entz & Latif 1974), but these features need more detailed investigations.

BIOLOGICAL PHENOMENA

For the settlement and fast development of lacustrine species in the reservoir, it is of high importance to remember that a part of Lake Nasser belonged earlier to Aswan Lake, the lake formed behind the Old Aswan Dam built at the beginning of the twentieth century and heightened in 1933. Although Aswan Lake was flushed year by year by the flood, the opportunity existed for the settlement of several limnophilic organisms. At present very important biological changes occur year by year and stabilized conditions can be expected only after the lake has reached its final working level.

Plankton

The phytoplankton (mainly diatoms and bluegreen algae) gives a fairly high primary production of 3.21–5.23 g $C/m^2/day$ gross production (Samaan 1972). Previous to 1972 water coloration caused by *Volvox* sp. appeared regularly in springtime in several localities. During the present years water coloration and extensive water blooms of Cyanophyta (e.g. *Microcystis aeruginosa*) became common annually for several months before

293

the flood period particularly in the southern areas of the lake between Wadi Halfa and Abu Simbel.

The abundance of zooplankton shows an annual cycle. The peak of its development is not simultaneous all over the lake but shows a longitudinal displacement in time towards the High Dam. As the front of the flood water passes the lake from July to December from the Second Cataract to Aswan, the richest zooplankton development could be noticed always just in front of the flood, in areas not yet affected by the flood water. The highest peaks could be observed in the central sections between 150 and 250 km from the HD. Zooplankton was the poorest usually in the first wave of the flood waters. The khors may play an important role in the repopulation of plankton in the new water masses brought by the flood. The flood water does not intrude into the inner parts of the khors where the rich plankton population remains untouched. The total biomass of the zooplankton was estimated roughly as 100,000 tons (fresh weight).

Littoral areas

The abundance of periphyton all over the littoral zone plays an important role as a nutritive basis for fish either directly for *Tilapia* or indirectly through insects or crustaceans (*Lates*, tiger fish etc.).

As is usual in newly formed reservoirs chironomids developed in enormous numbers. Their larvae were very common down to 3 to 5 m depth during the period 1969–1972. Although chironomid swarms were common on the shores in the evening hours almost all over the year, their peak numbers showed a similar timetable as the zooplankton, appearing always in front of the turbid flood, in water of high conductivity. Their larvae became important as fish food together with water bugs (Coryxidae) similarly widespread in the shallow littoral zone. At that time the biomass of chironomids was estimated around 100,000 tons in the lake. During these years the rocks of the shoreline became covered on their wind protected side by billions of dead chironomids (imagines). This led to an outburst of development of spiders, dragonflies, flies and toads all feeding mostly on the larvae or adults of this insects. In 1973 and even more in 1974 the number of Chironomidae and Coryxidae has been reduced in Lake Nasser considerably. Consequently in 1974 this was followed by the reduction of spiders, flies and even toads around the Lake Nasser. Instead of chironomids snails and shrimps *(Caridina nilotica)* became widespread in Lake Nasser in shallow areas among submerged littoral vegetation (e.g. *Potamogeton pectinatus*, *Najas* spp.). This kind of 'succession' of the littoral fauna could be followed year by year in Lake Nubia in horizontal direction as a consequence of the increasing water level flooding each year new shore areas. But there these features appeared not so obvious as in Lake Nasser because of the semi-riverine or even riverine character of Lake Nubia.

294

In several localities a fresh water crab *Potamonautes niloticus* (?) appeared in Lake Nasser (see ch. 25). Several new fish speces invaded the area and there were indications even of crocodiles.

Benthos

Before the High Dam has been constructed, a very rich mollusc fauna (Bivalvia) covered the bottom. These mussels did not move out of the river bed when Lake Nasser was formed and died all during the stagnation period because of lack of oxygen. In this area only millions of empty shells are present in the uppermost laycr of the mud. The benthos of the old river bed in Lake Nasser is inhabited by a very dense population of *Tubifex* sp. and some scattered chironomids belonging to the '*plumosus*' group. The standing crop of *Tubifex* exceeds in the old river bed 60,000 tons. Benthos is practically absent in the Second Cataract area where the rapid sedimentation of silt prevented the settlement of organisms. In the gorge region of Lake Nubia the original river fauna is present under almost pure riverine conditions.

Starting from 1973 a resettlement of mussels and oligochaetes is going on shallow inlet water, where the oxygen conditions are favourable all year round.

It can be concluded that the reservoir is an eutrophic water body with very favourable physical, chemical and biological conditions. In this way the final development of a very rich flora and fauna can be expected, which gives excellent prospects for the lake fishery. The expected yield is 20,000 tons of fish or perhaps more annually.

Macrovegetation

During the first years of its existence, larger aquatic plants were practically absent from Lake Nasser. The first submerged specimens *(Potamogeton pectinatus)* have been reported in spring 1972 in 0 to 2 m deep water in sandy littoral areas. Most of these stands dried out within the low water period in summer, but recovered again even in larger extent during the next winter and spring. The same happened in the following years and at present *Potamogeton pectinatus* is widespread all over the littoral zone of Lake Nasser.

Later on other species appeared. In summer 1973 *Najas armata* and *Najas minor* could be detected mainly in the northern sections of the reservoir in 3 to 7 m deep water. (Täckholm 1974). They formed in some places in 1974 very dense stands especially near the High Dam. Floating buds or branches of all species mentioned above could be observed in summer 1974 in several open water areas and khors of the lake. All these stands became very rich in periphyton and animal life e.g. freshwater shrimps, insect larvae, fish fry etc. Gradually the macrovegetation may get more abundant and more important as food source especially for young fish in the lake.

Potamogeton crispus, one of the most common vascular plants in 'Aswan Lake,' the lake between the High Dam and the Old Dam, did not spread until now in Lake Nasser although some few specimens were already present in the lake for years, but only in the immediate neighbourhood of the High Dam.

Another type of vegetation is formed in the most remote areas of the khors. Here in very shallow sandy places, very dense stands of *Chara* sp. are present. In shallow localities scattered weak stands of the reed, *Phragmites australis*, are located; probably the extremely strong fluctuation of the water level hindering their massive expansion.

Up to the present there are no floating weeds living in the reservoir, although some of them (e.g. *Eichhornia crassipes*, *Pistia stratiotes*) play an important but mostly disadvantageous role in some sections of the Nile River system both upstreams and downstreams. Consequently their appearance is possible in the future and has to be watched.

LAKE SHORE VEGETATION

During the first decade of its existance (1964 to 1974) the lake shore vegetation of Lake Nasser showed four successive stages in its development:

1. Inundation and disappearance of the previous riverside vegetation (1964 to 1968).

2. Temporary appearance of the first settling plants (1969 to 1970).

3. Development of semipermanent or permanent flora between the maximum and the actual water level composed by elements of different origin: A. Desert flora; B. Plants (weeds) from cultivated land and C. Riverside flora (1971 to 1973).

4. Progress of riverside flora around Lake Nasser (1973 to 1974).

The conditions around Lake Nubia were slighly different, because of the slow expansion of the lake above the Second Cataract. In this area there is a gradual withdrawal of the old flora towards the south, which is still going on, and a mixture of old and new flora elements can be observed frequently.

During stage 1. the ancient riverside vegetation (including 500,000 date palm trees) disappeared around Lake Nasser and the shore became completely bare. Seedlings appeared firstly during stage 2. in windprotected bays scattered around the main channel of the lake. Millions of seedlings of *Glinus lotoides* and *Tamarix nilotica*, the two dominant species at that time, were usually arranged in rows following stepwise the borderline of the decreasing water level after the flood. In this process water transport of the seeds may have played an important role.

During the third stage a rapid development of the plant cover could be

296

detected. Beside the two above mentioned species, a third dominant element *Hyoscyamus muticus* enriched the vegetation along the main channel of the lake. Simultaneously several other species appeared. Among the newcomers there were representatives of the vegetation from the previous wadis. These formed often a belt of small bushy plants just at the line of the previous highest lake level (*Fagonia bruguiera*, Cruciferae, Compositae, *Citrullus colocynthis* etc.). These species, except *Hyoscyamus muticus*, were most widespread around the landlocked ends of the khors.

Another spreading centre was in the neighbourhood of the High Dam between the highest and the present water level, where a flora, the richest in species, developed with several mainly bushy plants like *Zygophyllum coccineum*, *Rumex dentatus*, *Rumex vesicarius*, *Phragmites australis* etc. Some of them reached a height over 2 or 3 meters. The vegetation showed in several areas a surprising beauty and transformed the empty desert within 2 or 3 years into a 'park', (e.g. in the surrounding of Khor El Birba, 10 km south from the High Dam of Aswan). Not only sandy places but even stony and rocky areas became covered by dense vegetation during this stage.

It is suggested that the spreading of seeds was favoured by three main factors: a. the prevailing winds, blowing from the northern Nile valley along the river, b. the flood water reaching the area from the south and c. biotic factors such as man and animals, reaching the lake from different directions.

As the water level became more constant, like in 1973–1974 in stage four, the number of some previously predominant species became reduced remarkably (e.g. *Hyoscyamus muticus*, *Glinus lotoides*) and other species became widespread like *Portulacca oleracea*, different grass species etc. These changes may have been promoted by the appearance and very strong development of land insects, especially Orthoptera, grazing on the leaves of the above mentioned bushy plants (*Rumex* spp., *Hyoscyamus muticus* etc.). During this period, especially in summertime, different grasses and *Tamarix nilotica* appeared in large numbers forming dark green spots mainly on flat beeches. Some of the plants mentioned are able to withstand a shorter or longer period of submergence and recover after the flood (e.g. *Tamarix* spp.) but others die very shortly after being flooded (e.g. typical desert plants).

As the lake level has not yet reached its final working level, the final settlement of the new shore flore will take time.

REFERENCES

Aswan Meteorological Reports. 1926–1955. Cit. op. Omar & El Bakry 1970.

Entz, B. A. G. & Latif, A. F. A. 1974. Reports on surveys to Lake Nasser and Lake Nubia 1972–1973. Lake Nasser Dev. Centre Working Paper No. 6, Aswan pp. 137 + 14.

Hammerton, D. 1969. Blue Nile Survey. Hydrobiological Research Unit of the University of Khartoum Ann. Rep. 15, 1967–68, 5–13.

Hutchinson, G. E. 1957. A treatise on limnology. vol. I. p. XIV + 1015. John Wiley and Sons, Inc.

Maucha, R. 1948. Einige Gedanken zur Frage des Nährstoffhaushalts der Gewässer. Hydrobiologia 1: 225–237.

Nessim, R. B. 1972. Limnological study of Lake Nasser. MSC. Thesis, Alexandria Univ. Egypt pp. 1–295.

Omar, M. H. & El-Bakry, M. M. 1970. Estimation of evaporation from Lake Nasser. Meteorological Research Bulletin 2, pp. 1–27.

Ruttner, F. 1931. Hydrographische und hydrochemische Beobachtungen auf Java, Sumatra und Bali. Arch. f. Hydrobiol. Suppl. 8: 197–454.

Ruttner, F. 1962. Grundriss der Limnologie 3. Aufl. Walter de Gruyter, Berlin. p. 1–332.

Samaan, A. A. 1972. Report on the trips of Lake Nasser to investigate its primary production during March 1971. (unpublished).

Täckholm, V. 1974. Students' flora of Egypt. 2nd Ed. Cairo Univ. edition, Beirut Coop. Print. Comp. p. 1–888.

Williams, T. R. 1970. The river crabs of the Sudan. Hydrobiological Research Unit of the University of Khartoum. Ann. Rep. 16, 1968–69, pp. 10–14. (See also ch. 25, article by T. R. Williams).

19b. FISHES AND FISHERIES OF LAKE NASSER

by

A. F. A. LATIF

EDITOR. Dr. Latif's contribution has been shortened; part of it overlaps with the chapter by Dr. B. Entz and his observations on the distribution, behaviour and biology of fishes, although of documentary value for the present phase, will undergo considerable changes with the evolution of stable changes in this new reservoir. Some changes in the nomenclature of species have been suggested by Dr. P. H. Greenwood.

Fish Species

The fishes recorded in the Lake are known to the River Nile system and originated from those formerly living in the Nubian part of the Nile before impoundment. The 57 species recorded since 1964 belong to 15 families as listed below (Latif 1974):

Table 1. List of fish species recorded in Lake Nasser.

Family	Species	New nomenclature
1. Protopteridae	*Protopterus aethiopicus*	
2. Polypteridae	*Polypterus bichir*	
3. Mormyridae	*Mormyrops anguilloides*	
	Petrocephalus bane	
	Marcusenius isidori	*Pollymirus isidori*
	Gnathonemus cyprinoides	*Marcusenius cyprinoides*
	Mormyrus Kannume	
	Mormyrus caschive	
	Hypeopisus bebe	
4. Gymnarchidae	*Gymnarchus niloticus*	
5. Characidae	*Hydrocynus forskalii*	
	Hydrocynus lineatus	*H. vittatus*
	Hydrocynus brevis	
	Alestes nurse	
	Alestes baremose	
	Alestes dentex	
6. Citharinidae	*Distichodus niloticus*	
	Citharinus citharus	
	Citharinus latus	
7. Cyprinidae	*Chelaethiops bibie*	
	Barilius niloticus	
	Barilius loati	

ating greatly over *T. nilotica*, is in favour of fisheries of these species as the former is prolific, producing more eggs.

The future of *Lates niloticus*, as one of the predatory fishes and as the best fish for consumption, has to be considered, The landings of this fish increased from about 5 m.t. in 1966 to about 583 m.t. in 1971 but declined afterwards. Fishing is concentrated upon age groups II and III which constitute 60% of individuals landed. Based upon catch-curves, survival rate was found to be around 25% of the fish population. This together with low flood may have accounted for lower catches of 1972 and 1973 (Latif & Khallaf 1974 and 1975).

On the whole, to achieve the predicted potential, both improvement in the fish collection system and diversification of the fishery will be needed. The exploitation of the less valued species, such as 'kalb' *(Hydrocynus)* and 'raya' *(Alestes)* has to be considerably increased relative to the *Tilapia*.

Fish processing

Some of the fishes of Lake Nasser are salted. The amount in the first years (1966–1970) comprised on the minimum 40% and on the maximum 53% of total fish production. In the last four years, this value is lower up to a level of about 24% in 1973 but is mostly around 35%. The four prominent fishes salted are 'raya' (*Alestes* spp.), 'kalb' (*Hydrocynus* spp.), 'lebis' (*Labeo* spp.) and 'shilba' (*Eutropius* and *Schilbe* spp.). The cleaned gutted (for large fishes) fishes are rubbed in salt and packed in tins with salt in between layers and on the surface. The grade of the product varies, being of the highest grade for *Alestes* and the lowest for *Eutropius* and *Schilbe* spp. Salted fish find market mostly in upper Egypt but with improved methods the product could be more popular.

Sun-dried fish is limited but the possibility for this type of product is high due to the high temperature in Aswan Area. *Tilapia* spp. gave the best results but *Lates* and *Labeo* gave product of inferior quality.

Experimentally it was proved that *Alestes* and *Hydrocynus* spp. could be smoked and canned and the product has good quality. However, this method could not be applied on a commercial scale before understanding the fishing possibilities for these fishes from the lake, especially in view of the changes in the distribution of *Alestes* spp. with season and under different flood levels and conditions.

Fishing gears

Gill- and trammel-nets are commonly used for fishing in the lake. The former catch mainly *Alestes* and *Hydrocynus* spp. in Lake Nubia. The main fishes caught by trammel nets are Nile 'bolti', Nile perch and cyprinids. Formerly, these nets were made of cotton and deteriorated within two or three months by bacterial action under the hot summer conditions. Synthetic fibres are introduced and nets used are of 2–3 years longevity and not more than 3 m deep and have better catchability. Experimental deep gill nets, 10–12 m deep, made of mono- or multifilament twine gave good results with the former superior in production.

All fishing experiments and most of the commercial fishing are undertaken in the surface waters. Application of sunken gill-nets with wide mesh in the deep waters is limited mostly to the winter season, when the whole water mass is oxygenated, and is mostly used for fishing large *Labeo*, *Clarias* and *Bagrus* spp. Drifting gill-nets gave good catch of *Lates* and *Bagrus* in khor Kalabsha. Hoop nets, fixed in areas 10 m deep, gave good catch of *Tilapia* spp. in winter from the same area. Such a method when proved applicable on a commercial scale could be a breakthrough for increasing the catch during the present low fishing season and for subsequent management of the fisheries by prohibiting fishing during the spawning peak of *Tilapia* in March/April without minimizing the annual production. Furthermore, fishing in open waters has been till now limited and fishing possibilities from these vast areas will be investigated in the future.

Fishing boats

The activities of Lake Nasser fisherman are restricted by the fact that their only means of propulsion is the rowing boat, which may be suitable for fishing in the Nile but is not well-adapted to lake conditions. Small flat-bottom canoe type Alexandrian boats are predominant in the northern part. They can be manned by two fishermen and are adequate for fishing near the shore. The traditional Nile River type is found in the southern part of the lake, is larger, heavier, broader and manned by four or five fishermen. They are not considered well-suited for lake conditions, as they are heavy to row for large distances whereby the radius of fishing will be limited.

Rowing fishing boats of new designs but with easier manoeuvering have to be used on a larger scale. Ferro-cement mechanized boats are introduced and according to their size and design can serve different purposes of fishing and/or transportation. It will take some time until this type of boat is proved practical for a wide application and until the fishermen become convinced by this new design. The use of out-board motors has begun recently under FFHC (Freedom from Hunger Campaign) Project but progress of utilization of these motors is slack due to shortage of well-trained mechanics for maintenance and due to the belief by some fishermen that sound of these motors makes the fish escape. Undoubtedly, for wider use of longer nets mechanized method will be the only way.

Since the appearance of Lake Nasser, the number of fishermen gradually increased to about 5,000 for 1974 as compared with about 3,500 in 1970. They originally came from the Southern governorates of Upper Egypt and particularly Sohag and Qena. They pass most of the time in the Lake but their number decreases in the low fishing winter season. Not all the fishermen are members of Fishermen's Cooperative Society whose membership is based on ownership of fishing boats and gears and this society has only about 1,500 members.

Fishermen are living under difficult conditions and pass most of their time on the shore and on their boats. Sociological studies revealed the necessity of establishing ten villages – each with 300 houses – along the shores of the lake. Settlement plans are under high consideration and a government responsibility.

REFERENCES

Azim, M. E. A. 1974. Biological studies on *Tilapia nilotica* L. and *Tilapia galilaea* Art. in Lake Nasser. M.Sc. Thesis, Faculty of Science, Alexandria University, A.R.E.

Bazigos, G. P. 1972. The yield pattern of Lake Nasser (Arab Republic of Egypt). FAO Rome, St. S./1, September, 1972.

Ben-Tuvia, A. 1960. The biology of the cichlid fishes of Lakes Tiberias nad Huleh. Bull. Res. Coun. Israel Sect. B., Zool., 8b, 123–88.

Blache, J., Miton, F., Stauch, A. Iltis, A. & Loubens, G. 1964. Les poissons du bassin du Tchad et du bassin adjacent du Mayo Kebbi. Mem. Off. Rech. Scient. Tech. Outre-Mer, no. 4.

El-Bolock, A. R. & Koura, R. 1961. The age and growth of *Tilapia galilaea* Art., *T. nilotica* L. and *T. zillii* Gerv. from Beteha area (Syrian Region). Hydrobiol. Dep., Inst. Freshwater Biol., Cairo, Notes and Memoirs, no. 59, 27 pp.

El-Zarka, S., Shaheen, A. H. & Aleem, A. A. 1970. *Tilapia* fisheries in Lake Mariut. Age and Growth of *Tilapia nilotica* L. Bull. Inst. Ocean. and Fishes., Arab. Rep. Egypt, 1, pp. 149–182.

Entz, B. 1972. Comparison of physical and chemical environmental conditions as background for freshwater production in Volta Lake and Lake Nasser. Proc. IBP-UNESCO Symp. productivity Problems of Freshwaters, Poland, May 6–12 1970, pp. 883–91.

Entz, B. 1973. Morphometry of Lake Nasser. Working Paper no. 2, Lake Nasser Development Centre, Aswan, A.R.E.

Henderson, H. F. 1973. Actual and potential yield of fish in Lake Nasser, A.R.E. Rept., Lake Nasser Developm. Centre ROME.

Jensen, K. W. 1957. Determination of age and growth of *Tilapia nilotica* L., *T. gallilaea* Art., *T. zillii* Gerv. and *Lates niloticus* C. et V. by means of their scales. K. norske vidensh Selsk. Forh., 30, 150–157.

Latif, A. F. A. 1974a. Fisheries of Lake Nasser. Aswan Regional Planning, LNDC, A.R.E., 235 pp.

Latif, A. E. A. 1974b. Fisheries of Lake Nasser and Lake Nubia. In report on: Trips to Lake Nasser and Lake Nubia. By: B. Entz & A. F. A. Latif; Aswan Regional Planning, LNDC, A.R.E.

Latif, A. F. & Khallaf, E. A. 1974. Studies on Nile perch, *Lates niloticus* L. from Lake Nasser. Bull. Inst. Ocean. & Fisher. A.R.E. 4, pp. 131–163.

Latif, A. F. A. 1975. Population study on Nile perch. Hydrobiologia (in press).

Latif, A. F. A. & Rashid, M. M. 1972. Studies on *Tilapia nilotica* from Lake Nasser. I. Macroscopic characters of gonads. Bull. Inst. Ocean. & Fisher., A.R.E., 2, pp. 215–238.

Mahdi, M. M. 1972. Factors affecting the survival and distribution of some Nile fishes. Ph.D. Thesis, Faculty of Science, University of Cairo.

Ryder, R. A. 1973. Fish yield projections on the Nasser reservoir, A.R.E. (including Lake Nubia, Sudan). FAO, Rome.

Samaan, A. A. 1972. Report on the trip of Lake Nasser to investigate its primary productivity during March, 1971. Rept. for LNDC, A.R.E.

307

20. DELTA LAKES OF EGYPT

by

J. RZÓSKA

Along the coast of the Egyptian Delta extend four large lakes, all in some connection with the Mediterranean Sea. Their origin is probably nilotic, some lie at or near the mouths of the ancient outlets of the Nile and all lie at an altitude of 0–30 m a.s.l. From west to east they are: Lake Maryut (Mareotis), Edku, Borullus and Menzalah. Some essential data are:

	Area km²	Salinity	Depth m	Biology
Maryut	260	var. 3–5% Cl	1 m	Locally polluted
Edku	131	0.4% Cl	1.5 (max.)	Very differentiated
Borullous	546	0.9–2.8	0.5–2.0	Brackish
Menzalah	2360	0.8–2.8	0–1.20	Mainly brackish

Sandbars, some of the consolidated, separate the lakes from the sea but during storms strongly saline water can penetrate into deeper parts of the lake areas.

Geological and Hydrological investigations

Geological and hydrological investigations have revealed that the Delta forms a huge underground reservoir, with the Nile as main re-charger. The enormous sediments accumulated during many thousands of years have two water bearing strata, the upper composed of alluvial clay and silt, saturated with fresh water, and a lower stratum of sands and gravel; a salt water layer lies underneath the freshwater sediments. The whole 'aquifer' area is inclined north with a gradient of 0.75 m/km the freshwater area of underground water does not occupy the whole delta, the northern part has a saline groundwater. Only half of the Delta is intensively cultivated (Rofail & Tadros, Shata & Fayoumy 1969).

Investigations of the coastal area of the Delta have been carried out for a number of years; Steuer carried out hydrobiological research in the 'foci del Nilo' in 1933 and incorporated a great deal of previous knowledge (Steuer 1942). A number of publications appeared from the Marine Laboratory at Alexandria, now the Alexandrian Institute of Hydrobiology, other work was done by universities; finally the inter-national organisations of FAO and UNESCO organised substantial further investigations. The FAO project 1954–1958 concentrated on fisheries and the bases of biological productivity of the Lakes with a series of important publications in the form of FAO reports or papers in journals. Of these the most comprehensive is the 'Beiträge zur Limnologie Ägyptens' by Elster & Vollenweider (1961) and other papers. Only a summary of the main data is presented here briefly.

1. The existence of the lakes in their present character depends on the steady inflow of Nile water especially during the flood. This fresh water is supplied mainly by drainage canals which receive surplus water from irrigation. This water keeps the balance between fresh and saline character of these lakes.

2. This Nile water supplies also nutrients with the Nile silt. In a special investigation Elster and Gorgy (1959) have proved without any doubt that the Nile silt is rich in nutrients; Jannasch (1957) has found that the bacterial flora of the flood water finally sediments out with the silt particles with a subsequent release of nutrients.

3. All the coastal lakes are shallow, with low shores and with very abundant development of *Phragmites* 'forests' with open water clearings inside, providing feeding grounds and shelter for a very rich fish population. Other plants like *Potamogeton, Ceratophyllum, Najas* form in parts of the lakes underground meadows of considerable extent. In Edku lake *Eichhornia* appears in masses in summer but vanishes in winter, when the water temperatures may fall to 11–13 °C.

4. Limnological characteristics can be very variable; oxygen may form often steep gradients even in the shallow water, it may come to de-oxygenations, but transient and on the whole short. H_2S develops often especially in reeds but does not seem to persist long enough to prevent enormous shoals of young fishes from living there. Salinity is mainly of the oligohaline degree but may increase both locally and temporary. A chemocline has been observed on several occasions, again not permanent. Transparency is not high (up to 0.35 m), the water is often turbid especially during winds and near the inflow inlets of drainage canals; pH varies between 7 and 10.

5. In very general terms, the peculiar nature and layering of 'brackishness' may cause the plankton to be mainly of freshwater composition, while the bottom fauna is ver often decidedly brackish or marine.

6. The lakes are biologically rich both in plankton and benthos; Vollenweider (in Elster & Vollenweider 1961) calculated the production of phytoplankton in biomass (Maryut) as 50,000 to 180,000 kg/ha/year and by C 14 at 1g $C/m^2/day$ and much higher values; macrophyte production was estimated in the growing season at 0.75–2.55 g $C/m^2/day$. Enormous figures for the benthos are quoted for some sites, e.g. in L. Menzalah: 54,000 *Gammarids* and *Corophium*, 6,700 *Leander*, 3,500 Hydridae, 200 molluscs per m^2 not counting polychaets and balanids.

This richness is reflected in the fishes and their yield. The fish fauna has predominantly nilotic character with some marine additions. According

310

to Boulenger (1907) 83 species are found in these lakes, composed of: Mormyridae 11 species, Characiodae 14, Cyprinidae 13, Siluridae 24, Cichlidae 5; Mugilidae 3, Serranidae 3 species. *Tilapia* appears in three species and is the most important component of catches, with 60–94%. Mugil is probably the second contributor to the yield. Yield per ha is much higher than in Europe and may reach 300 kg per year, a high figure for not-managed waters. Fishing is intensive, 26000 fishermen are active, often excessively without regard to conservation of stocks and disregarding existing regulations as to mesh sizes prescribed for the protection of young fishes. Although available statistics (1956) are unreliable, they show some approximate values of 80–100,000 ton of fish per year for the whole of Egypt. Of this total about 25% came from the sea, 15% from the Nile and canals, the rest 60% form the coastal lakes. The fertility of the lakes was based in the past on inundations, acting as breeding grounds, and on the fertilizing effect of Nile sediments brought in by flood waters.

This the past. The new hydrological regime brought about by the Aswan High Dam has cut completely the two factors of fertility mentioned above. There will be now a constant but smaller flow of Nile water into the lakes, totally devoid of sediment. No inundations will occur anymore. Further, a clash of interests between agriculture and fishery is evident; extensive land reclamation schemes point badly for the future of fisheries. Lake Borullus has already been investigated under the new regime of water supply (H. M. El Sedfy & J. Libosvarsky 1974) and a gradual decrease of fish production is expected. Elster (1961b) has given a brief but impressive survey and recommendations for the conservation of fisheries in this area including the introduction of fish farming, which is already being investigated by scientific pilot schemes. For the new fishing grounds along the new lake created by the Aswan High Dam see ch. 19.

REFERENCES

Boulenger, G. A. 1907. The Fishes of the Nile. In: Anderson, Zoology of Egypt, London.
Elster, H. J. & Gorgy, S. 1959. Der Nilschlamm als Nährstoffregulator im Nildelta. Naturwissenschaften 46: 147.
Elster, H. J. & Vollenweider, R. 1961a. Beiträge zur Limnologie Ägyptens. Arch. Hydrob. 57: 241–343.
Elster, H. J. 1961b. Binnenfischerei in Ägypten. Die Umschau in Wissenschaft und Technik 22: 681–684.
Jannasch, H. W. 1956. Vergleichende bakteriologische Untersuchungen der Adsorptionswirkung des Nil-Treibschlammes. Ber. Limnol. Fluss-station Freudental 7: 21–27.

Rofail, N. & Tadros, S. 1969. The study of modelling the ground water flow in the Nile Delta using the electrical Analogue Method. Hydrology of deltas IASH/AIHS–Unesco, Paris, vol. 2: 408–415.

Hussein, M., El-Sedfy & Libosvarsky, J. 1974. Some effects of Aswan High Dam on water and Fishes of Lake Borullus, Zoologicky Listy 23: 61–70.

Shata, A. & El Fayoumy. 1969. Remarks on the regional geological structure of the Nile Delta. Hydrology of Deltas, vol. 1: 189–197. IASH/AIHS–Unesco.

Steuer, A. 1942. Richerche Idrobiologiche alle Foci del Nilo. Mem. Istituto Ital. Idrobiol. 1: 85–106.

There is a considerable number of special papers on algae and zooplankton, some of these are contained in the bibliography in Elster & Vollenweider 1911a.

V. HYDROBIOLOGY AND LIMNOLOGY
OF THE WHOLE RIVER SYSTEM

21. THE INVASION OF EICHHORNIA CRASSIPES IN THE SUDANESE WHITE NILE

by

J. RZÓSKA

From the entry into the Upper Nile swamps downstream to the Gebel Aulia Dam the great invasion by *Eichhornia crassipes*, probably in 1957/58, has become a major event. It is at present impossible to assess whether and what lasting biological changes will occur. The previous chapters written on this river region are based mainly on investigations depicting conditions which existed up to 1956. They are left as documentary record against future developments.

History of discovery and spread in the Nile

On the 27th March 1958 P. A. Gay, a member of the Hydrobiological Research Unit, University of Khartoum, found the plant first at Aba Island and further south over a river course of 1,000 km. (Gay 1958); this was confirmed in April 1958 (Wickrama Sekara 1958). The infestations were serious enough to inform the appropriate authorities. Unfortunately, no immediate action was taken.

On enquiry, Gay found that the plant was already seen at Kosti in December 1957. Wickrama Sekara, travelling up the Nile in August 1957, although familiar with the water hyacinth from infestations in Ceylon, did not see any signs of infestations; maybe single plants were present on the banks. Gay (1956/7, 1957/8), who made an investigation of the riverain flora of the White Nile in 1956/7, found no trace of the plant in 360 photographs. Nor was any infestation seen during the journeys of the Hydrobiological Research Unit in any part of the B. el Gebel section of the White Nile prior to 1956. It is not excluded that the plant had already arrived in remote backwaters which are difficult to access.

In August 1958 plants appeared near the Gebel Aulia Dam; in October 1958 the whole river from Juba to the Dam over a distance of 1800 km was infested. In January 1960 and December 1961 the infested area was estimated by two members of the University (Gayed 1961/63). Over a stretch of about 1600 km of the White Nile from Kosti to Juba an approximate area of about 1240 ha was occupied by floating mats. Further observations were made in the autumn of 1969, December 1970 and June/July 1971 indicating somewhat less infestation in the northern stretches (Faris F. Bebawi 1973).

In 1960 government agencies became aware of the seriousness of the situation and control measures were taken preceded by aerial assessment. These continue up to the present. At times a staff of 200 was employed at a cost of 500,000 Sudanese pounds per year (Heinen *et al.* 1964). In spite of these costly operations, the present situation remains as it was in 1960 according to Obeid (1974).

CENTRES OF INFESTATION

It seems that the Upper Nile swamps have become a 'perennial' centre of infestation, from where a continuous supply of plants go north with the current. At present almost all the swamp rivers with their adjacent lagoons are occupied by the water hyacinth with the exception of the Bahr el Ghazal where only the first 70 km of its 200 km course have a low infestation with mainly stunted plants. The upper part of this river has special limnological characteristics of, e.g. low pH, and biologically is known for its wealth of Desmids (see chapter on Upper Nile swamps). A. Berg (1959) reported that in the Congo (Zaire) and Rwanda *Eichhornia* does not invade waters of pH 4.2. No such low pH has been recorded in the upper Ghazal.

Agents of spread in the Nile are currents and winds. Currents in the swamps are subject to flood conditions in the White Nile governed by the rainy season in Uganda and locally. However, all swamp rivers meander, are hemmed in by vegetation, especially *Vossia*; this plant makes forays into the river and creates small bays and niches which favour the establishment of *Eichhornia crassipes*. Here it may be mentioned that another species, *E. natans* Solms, was a permanent member of the swamp flora but never became conspicuous.

Once the river leaves the swamps it runs mainly northwards. Its current is slowed down by a very low slope and by the stowing action of the Gebel Aulia Dam, noticeable upstream for about 500 km. The lake-like conditions of the dam basin are favourable for the water hyacinth. Floating islets of different composition have been known in this region for a century and many observations exist on blockages ('sudd') which torn-off parts of the swamp vegetation have caused (see State of River). Such rafts rarely passed the river port of Kosti. *Eichhornia* rafts, because of their compactness, seem to travel much further north than the lighter rafts of before.

Winds in the Sudan have two main directions; they come during the 'summer' months about April until October from the south, and from November until March from the north. These alternating winds drive the water hyacinth north in the summer and disperses it southwards toward the swamps in winter. So far, the northern limit of the *Eichhornia* spread has been the masonry of the Gebel Aulia Dam. At times a vast and solid mat piles up against the dam, so thick that people could walk on it.

After control measures, large amounts sank down to the bottom near the dam, which is about 10–12 m deep. The situation creats problems for the operation of the dam which is closely coordinated with the regime of the Blue Nile where the Sennar Dam stores up water for the Sudan's greatest agricultural, cotton growing, Gezira scheme. A fast network of irrigation channels extends between the two rivers as seen even from a space photo (Fig. 64). The Blue Nile is at present not affected by *Eichhornia* but it would be a major calamity if the plant could penetrate via the pump schemes on the White Nile. Photographs supplied by M. Obeid show the plant piling up against the pump inlets.

PROPAGATION OF THE WATER HYACINTH

The extraordinary growth rate, estimated by Gay (1960) as about 10%–15% of its fresh weight per day, adds to the explosive propagation capacity, vegetative or by seeds. Obeid (1962) found that two plants isolated in a water basin increased to 30 in two and to 130 in three months. In his contribution (1974) to this chapter he quotes figures from Zaire (Congo) where two plants increased to 482 in two months and 1,200 in four months; a hyacinth mass of 150 tons per hour was passing Leopoldville in the Zaire river 'despite the expenditure of 50 million francs a year . . . to keep the river clear'.

A single plant sends out tillers (off-shoots) which may bind single plants into rafts. Some of these have an area of several hundred square metres and can coalesce into continuous mats. Breaking up of rafts either by natural agencies or by river traffic only causes the creation of new nuclei of development. With a falling river, when the Gebel Aulia Dam is opened, many rafts get stranded along the shores, shrivel up but have a great capacity of recovery. *Eichhornia* flowers in hot conditions which will increase difficulties of control. Seedlings have been recorded by S. el Din Hassan Ahmed in 1962 and published by Pettet (1964).

Control measures in the Sudan

Mechanical control can only be used where labour is available and large parts of the White Nile river region are very thinly populated.

The hormone herbicide 2.4-D was applied by spraying from boats and planes but such operations are prohibited during the cotton growing season (July–December) and in the neighbourhood of villages. Young plants are killed more easily than mature ones. The campaign, started in 1960, is still going on – a sign of its difficulty. The infested Upper Nile swamps cannot be treated easily and from there the supply of plants to the vital northern stretch of the White Nile is continuous. Burning of stranded or raked out plant masses is now gradually recognised as necessary, where it can be applied.

The effects of the *Eichhornia* invasion affect navigation, irrigation pumps, fisheries, water supply to riverain people and water loss by increased evaporation. It may cause an increase in disease vectors (see ch. 22). The White Nile is an international and national waterway between the Sudan and Uganda, and an important link between the southern Sudan and the centre of the country at Khartoum. Steamers, usually with barges tied in front and on the sides, had to stop frequently to remove the plants wedged in between; delays and costs have been considerable. Irrigation pump schemes on the White Nile suffered because of choking of pumps. Fisheries are impeded; specific observations exist so far only from Shilluk country, immediately north of the swamps and on the Sobat River (Davies 1959), where a broad band of water hyacinth on the shores prevents people using their throwing nets. The same applies to people fetching their water supply from the river. (For limnological effects, see chapter 13).

There is no doubt that water loss by evaporation may increase; experiments on evapotranspiration were done in Egypt by Hammouda (1968). According to Hammouda the transpiring area of the water hyacinth increases 15 times with the growth of the plant from 5 to 45 cm of length and evapotranspiration was found in this study to be much higher than from free water. Even during the night stomata do not close completely.

In the Sudan the continuous arrival of plants from the south has created 'new swamps' in the Gebel Aulia basin especially around the large number of partly drowned trees (Gay 1959). The spread of snail vectors of bilharzia and of mosquitoes has to be watched.

A widespread publicity campaign has been launched to inform the population on the pest and legislation has been introduced against possession and transport of plants.

According to Obeid (1974) present research in the Sudan is centred on the temporal and spatial distribution of the plant, its main propagation foci and the role of river traffic so as 'to adopt a strategy of control . . . most effective and environmentally less hazardous'. As 'complete eradication of the water hyacinth is almost impossible . . . a state of equilibrium, ecological, economic and end environmental could be reached with well founded research and control programme' according to Obeid.

It may be worth while to follow up the brief investigations undertaken by Mitchell and Thomas on behalf of UNESCO in the regions which are the original home of the hyacinth and other explosive plants, tropical South America (UNESCO Technical Paper No. 12, Paris 1972).

The water hyacinth has become an international problem and international effort is needed.

Origin of invasion in the Nile system

In Egypt *Eichhornia crassipes*, introduced apparently as an ornamental plant, has been recorded from 1912 onwards (a.o. in Lake Edku) and it is listed in the Flora of Egypt by V. Täckholm and Drar. The Egyptian authorities are closely watching any possible arrival of plants from the south, as shown by the bars and jetties across parts of the river at Cairo and, no doubt, elsewhere. The enormous stretch of the new Aswan basin is a potential danger point.

Controversial opinions have been expressed on the origin of the invasion. Uganda and Ethiopia are free; no plants have been seen in southern Egypt and the northern stretch of the Main Nile. Fais Fariz Bebawi in his M.Sc. Thesis suggests as a most probably entry the region of Shambe in the middle of the Upper Nile swamps. Some river and wetland connections exist from there westwards with the Zaire (Congo) system, where heavy infestations started in 1952.

REFERENCES

Berg, A. 1959. In: Bull. Agricole du Congo Belge et du Ruanda vol. 50: 365–394.

Davies, H. R. J. 1959. The effect of *Eichhornia crassipes* on the people of the Sobat and White Nile between Sobat and Kosti. Annual Report of the Hydrobiological Research Unit, University of Khartoum, 6: (1958–59), 26–9.

Faris, F. 1972. Studies on the ecology of *Eichhornia crassipes* (Mart.) Solms. in the Sudan. M.Sc. Thesis, University of Khartoum, (unpublished).

Gay, P. A. 1956–7. Some aspects of the riverain flora of the White Nile and the B. el Ghazal. Ann. Rep. Hydrob. Res. Unit no. 4: (1956–1957) 10–20.

Gay, P. A. 1957–8. The riverain flora of the Nile. Ann. Rep. Hydrob. Res. Unit no. 5: (1957–1958): 7–15.

Gay, P. A. 1958. *Eichhornia crassipes* in the Nile of the Sudan. Nature, Lond. 182: 538–9.

Gay, P. A. 1960. Ecological studies of *Eichhornia crassipes* Solms. in the Sudan. J. Ecol. 48: 188–9.

Gay, P. A. & Berry, L. 1959. The water hyacinth: a new problem on the Nile. Geog. J. 125: 89–91.

Gayed, S. K. 1963. Some observations on the distribution of the water hyacinth in the Nile between Kosti and Nimule. Ann. Rep. Hydrobiol. Res. Unit, no. 9/10 (1961–1963): 8–12.

Hammouda, M. A. 1968. The water outlay by *Eichhornia crassipes* and observations on the plant chemical control. Phyton, Austria 18: 97–106.

Heinen, B. T. & Ahmed, S. H. 1964. Water Hyacinth Control on the Nile River, Sudan. Information Production Center, Dept. of Agric., Khartoum.

Obeid, M. 1962. An investigation into the mineral nutrition of some common weed species in the Sudan. M.Sc. Thesis, University of Khartoum.

Obeid, M. 1974. The water hyacinth and the Nile. Typescript, unpublished, here incorporated.

Pettet, A. 1964. Seedlings of *Eichhornia crassipes*: a possible complication to control measures in the Sudan. Nature 201: 516–517.

Tackholm, V. & Drar, M. 1950. Flora of Egypt, vol. 11. Fouad I University Press, Cairo.

Wickrama-Sekara, G. V. 1958. Crop protection files, Ministry of Agriculture, Khartoum 1958.

22. SCHISTOSOMIASIS IN THE NILE BASIN

by

C. A. WRIGHT

Schistosomiasis (Bilharziasis) is a disease of man and other mammals, including domestic cattle, caused by parasitic flatworms of the genus *Schistosoma*. The adult worms live in the abdominal blood system of their hosts, usually in the veins surrounding the intestine or those around the bladder. Eggs laid by the parasites work their way through the wall of the intestine or bladder and are voided in the hosts' faeces or urine. If the eggs fall into water they hatch and the ciliated larva (miracidium) which emerges, swims freely but must locate and penetrate into a suitable freshwater snail within a few hours. In Africa the snail hosts belong to the family, Planorbidae, species of the genus *Biomphalaria* serving as hosts for the common parasite of the intestinal venous system *(S. mansoni)* and species of *Bulinus* for the parasite in the bladder wall *(S. haematobium)*. Within the snails several developmental stages occur with progressive multiplication, leading after about a month to production of large numbers of the final larval stage (cercaria). Several thousand fork-tailed cercariae may develop from a single original miracidium. The cercariae, like the miracidia, have a brief free-swimming life and must quickly enter a suitable final host such as man. Entry into the final host is by direct penetration through unbroken skin and infection may occur as a result of any kind of contact with water containing cercariae. Maturation of the parasites in the final hosts takes two to three months and the adult worms may survive for many years.

Because schistosome parasites were originally discovered in man in Egypt by Theodor Bilharz in 1851 and the life-cycles of *S. haematobium* and *S. mansoni* were worked out by Leiper in 1915 at El Margh near Cairo, Schistosomiasis has often been thought to have its origins in the Lower Nile valley. Some support for this idea was found in the mention of haematuria, the most obvious symptom of infection by *S. haematobium* in both the Kahun papyrus dated about 2,000 B.C. and the Ebers papyrus of about 1,500 B.C., also the demonstration of eggs in the tissues of mummies of the 20th dynasty (about 1,250–1,000 B.C.). However, it is more likely that the Schistosome parasites of man evolved together with their final hosts in East Africa during and perhaps before the Pleistocene era. The contemporary distribution of human schistosomiasis in the Nile drainage largely reflects the distribution of snails of the host genera, *Biomphalaria* and *Bulinus*, but the picture is slightly complicated by certain specific restrictions on intermediate host compatibility in *Schistosoma haematobium*.

On the southern shore of Lake Victoria in the Mwanza region and in the east around Kisumu the host for *Schistosoma mansoni* is probably mainly *Biomphalaria sudanica* in streams and seepage areas. The region immediately to the west of Lake Victoria is largely free from schistosomiasis because the low conductivity waters do not provide suitable habitats for the snail hosts. In the lake itself, the endemic *B. choanomphala* has been shown to transmit *S. mansoni* and the related lacustrine species, *B. smithi*, in Lake Edward and *B. stanleyi* in Lake Albert may also be implicated as hosts. *S. mansoni* occurs in a number of areas in Uganda but the prevalence is particularly high in West Nile District (reaching 90% in some localities) where the principal snail hosts are probably *B. pfeifferi* and *B. sudanica*. Further to the north the prevalence is high in Equatoria Province of Sudan but the situation in the swamp region is unclear. It is probable that the sparse human population in the area is insufficient to maintain a high level of transmission. From about 160 kilometres to the north of Malakal to well below Khartoum there is a highly infected area with prevalence rates ranging between 60% and 90%. Throughout this area and around the Blue Nile above Khartoum, particularly in the Gezira where *S. mansoni* is also common, both *Biomphalaria pfeifferi* and *B. sudanica* occur but the first of these two species is probably the more important host for the parasite. *B. pfeifferi* is also responsible for transmission of *S. mansoni* in the highland areas of Ethiopia around Lake Tana and the headwaters of the Blue Nile. In the Northern Province of Sudan *S. mansoni* transmission is limited to some of the irrigated areas where again *B. pfeifferi* appears to be the main snail host. *S. mansoni* in Egypt is restricted to the Delta region where it is transmitted by *Biomphalaria alexandrina*. The parasite has not been recorded south of Giza so far but developments from the building of the High Dam are being watched with care.

Schistosoma haematobium occurs on the southern edge of Lake Victoria where it is carried by *Bulinus nasutus*, a snail which usually occurs in temporary pools. On the east shore of the lake transmission is probably effected largely by the same species and the related *B. globasus*, an inhabitant of more permanent water bodies. The presence of *B. truncatus* has recently been demonstrated on the Kano Plain near Kisumu and this species, which is capable of acting as a host for the Mediterranean strain of *S. haematobium*, may present an additional problem for the future. As with *S. mansoni* there appears to be little or no *S. haematobium* to the immediate west of Lake Victoria. No transmission is known to occur in the lake itself nor in the other great lakes in Uganda although *Bulinus coulboisi* which occurs in Lake Albert has been reported to be susceptible to experimental infection with the parasite. There are few loci of *S. haematobium* in Uganda and the dominant bulinid snails in the country (*B. ugandae* and *B. forskali*) do not appear to be susceptible to infection. The parasite is said not to occur in Equatoria Province of Sudan and the situation in the swamp region is obscure. From Kosti northward the occurrence of *S. haematobium* increases. In the Gezira and other irrigation schemes right down to the Egyptian border, urinary schistosomiasis has a high prevalence but control measures in the irrigated areas have helped to improve the situation. Throughout this region the intermediate host is *B. truncatus*, a tetraploid species which probably has its origins in the Ethiopian highlands where a polyploid series of bulinid snails (diploid, tetraploid, hexaploid and octoploid) is known to occur. Despite this, there does not appear to be any significant transmission of *S. haematobium* in the region of the headwaters of the Blue Nile. *S. haematobium* occurs throughout Egypt where it is a serious public health problem; the snail host is *B. truncatus* and both parasite and host appear to have become established at some places in the Lake Nasser area.

In conclusion it should be mentioned that although *Schistosoma bovis* was originally described from cattle in Egypt, it is questionable as to whether transmission of the parasite occurs in that country. It is probable that the original infected cattle had come from Sudan and at Kosti on the White Nile there is an important focus of *S. bovis* transmission which causes serious disease and high mortality in cattle. The intermediate host at Kosti has been named as *Bulinus ugandae* but this identification should be treated with reservation since the snails show characteristics not usually associated with that species.

EDITOR. A detailed study of the distribution of Bulinus and Biomphalaria in mid-Sudan (Williams & Hunter 1968) links the northern advance of some vectors, named as *Bulinus ugandae*, *Bulinus truncatus* and *Biomphalaria sudanica*, with the spread of *Eichhornia crassipes* since 1958. The authors stress also the great difference between the Blue Nile with faster current, steep banks with no vegetation and no snails (except the Sennar basin) and the wide, slow flowing and partly swampy White Nile with large snail populations.

The geographic distribution of the genus *Bulinus* and the role of the Nile as link between the species to the north and south of the Sahara has been recently discussed by Berrie (1970). The situation of molluscan vectors of Schistosomiasis in the Gezira was the subject of a study by Markowski (1953). There is a number of older publications on the extent of this problem in the Sudan.

REFERENCES

Quoted in Editor's note.

Berrie, A. D. 1970. Snails, schistosomes and systematics: some problems concerning the genus *Bulinus*. In: Advances in Parasitology 8: 183–188.

Mahdi, M. A. & Amin, M. A. 1966. An attempt to control bilharziasis by fish. Hydrobiologia 28: 66–72.

Markowski, S. 1953. The distribution of the molluscan vectors of schistosomiasis in the Sennar area of the Sudan, and their invasion of the Gezira irrigation system. Ann. Trop. Med. and Paras. 47: 5–380.

Williams, S. N. & Hunter, P. J. 1968. The distribution of *Bulinus* and *Biomphalaria* in Khartoum and Blue Nile provinces. Bull. WHO, 39: 949–954.

23. INSECTS AS FACTOR IN GENERAL
AND HUMAN ECOLOGY IN THE SUDAN

by

J. RZÓSKA & D. J. LEWIS

There are two main reasons for including this special chapter in our volume on the biology of the Nile. The first is that insects by mass appearances reveal important aspects of the general ecological set up of the Nile system; secondly they affect human and animal life strongly.

During the first population census ever held in the Sudan in 1955/6 some important facts have been established: of the population of 10263000 recorded then, 92% lived in rural areas – of these 15% or 16% were nomadic and pastoral, but occupying more than half of the country; this figure does not include the three southern provinces where regular seasonal movements are governed mainly by the hydrological regime of the river system. The 15% of the population in the more northern areas are concentrated mainly in Kassala Province (5%), Khartoum (4%) and the North (5%); the remaining 1% lives in Kordofan, Dafur and Blue Nile provinces. This distribution is significant as it reflects to a large degree the existence of camel-owning tribes (K. J. Krótki '21 Facts about the Sudanese, First Population Census of Sudan 1955/6', published 1958). When the great work of the census was finished, the Philosophical Society of the Sudan held its 6th Annual Conference discussing the results and their significance (Report of the 6th Annual Conference, Phil. Soc. Sudan, University Khartoum, 16–17 January 1958). One of the difficulties of the census was estimating the nomadic population. Biologists present were able to help the economists and statisticians to understand the reasons for migrations as outcome of insect pests. 'It is remarkable how close the map of nomadic movements corresponds with the map . . . of Tabanids . . . According to a very rough calculation – which will, it is hoped, be improved on further investigations – as many as 50% of nomads move primarily because of flies, 30% primarily because of shortage of grazing and only 20% primarily because of shortage of water' ('21 Facts' see ref. above). The argument advanced by the biologists was entirely based on the work by D. J. Lewis discussed in detail below dealing with the insect groups involved.

Tabanidae

In a paper on 'Early Travellers Accounts of Surret Flies (Tabanidae) in the Anglo-Egyptian Sudan', Lewis (1952) has summarised a large number

325

of older references from 1790–1900 on the effects of mass appearances of Tabanids on the state of the country and especially on the forced migrations of camel- and cattle-owning tribes of a large zone of the Sudan from the Abyssinian border across to the western part of the country. 'Although many of the reports are incorrect . . . all present an aspect of local history . . . (and) show, in the light of modern knowledge, that a considerable amount of general information about Tabanids existed in 1900' (D. J. Lewis 1952).

This was one of the first years after the occupation of the Sudan in 1898, following the 18 years of the Mahdi's uprising (1881–1898), which made the country inaccessible to travel and investigation. Intense activity of the new established Anglo-Egyptian authority started almost immediately by 'stock taking' of many aspects of the vast country (Gleichen 1905). 'Disease ridden' and 'insect infested', as early travellers described the country, it certainly was; it demanded years of investigations by many people to arrive at the scientific assessment of extent, cause and remedy of insect infestations. The results of this work on Tabanids is brought together by Lewis (1953b) in a concise paper listing 139 bibliographic references. Seventy species of Tabanids were recorded in the Sudan at that time, one or more species are probably involved in transmitting camel- and cattle-trypanosomiasis and human loiasis. The morphology and distribution of many of the 70 species are described and mapped. The distribution shows two main trends: a. many species are concentrated on rivers; b. they all fall into a broad distribution belt across the country from about 8° to 13° lat. N (Fig. 26 in Lewis 1953b).

The arid regions of the north are devoid of Tabanids except some few occurrences on the Nile in Nubia. Fig. 27 of Lewis's paper shows the belt with camel trypanosomiasis and also further south the tsetse area and the occurrence of human loiasis. Most illuminating for this chapter is the map, reproduced here (Fig. 81) showing the annual migrations of tribes owning camels or cattle. During the rainy season, in these latitudes from May to September, people and stock move from near the rivers and clay soils turning often into swamps, to more northern, drier sandy regions where the seasonal rains allow some grass to grow (see the map of isohyets across the country gradually diminishing towards the north, and the vegetation map – Fig. 10, 11) Some of these migrations are of great extent, e.g. from Khor Yabus to Manaqil of more than 500 km. Historically, during their conquest of the Sudan in the 13th to the 15th centuries, the Arabs were halted ultimately by the zone 'of biting flies and other adverse conditions' (Lewis 1953 p. 203 quoting MacMichael 1922). Now the camel-owning tribes live between 13° to 18° lat. N because the fly-borne camel disease kills camels south of 13° and lack of rain and grazing keeps them south of 18° lat. N. Cattle are less affected though *Trypanosoma evansi* may occur; it has been said that of the three million cattle in the Sudan in 1947 almost all have to migrate.

326

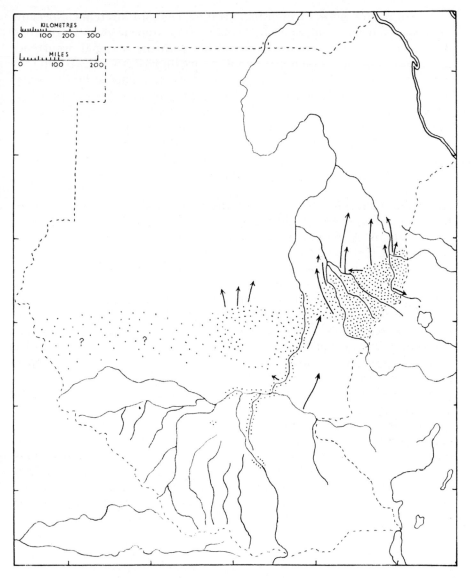

Fig. 81. Distribution of Tabanid flies and cattle migrations under their influence in the Sudan; from Lewis 1952.

In the southern Sudan, until recently a cattle rich territory, the main cause of migrations is the hydrological regime with flooding of plains, but biting flies also impose severe conditions on man and beast (see chapter on Upper Nile swamps). Wild animals ('game') also have to migrate and Baker (1871) and more recent observers have described these movements, largely caused by Tabanids.

Tabanids are permanent inhabitants of riverain stretches especially between Gebelein and Shambe. Inland they appear over large areas during the rainy season closely connected with the seasonal occurrence and amount of rain. The breeding habits are imperfectly known and only for some species; moist soil or nearness of water seems to be necessary. There may be only one generation per year; some species live probably several months as larvae, and hibernate or aestivate in this stage with the pupal stage apparently very short. Many factors causing the mass appearances are still unknown.

Simuliidae

According to Lewis (1953a, c) who summarised all the evidences at the time of writing, there are 22 species and 2 varieties of Simulium present in the Sudan. As habitat they require running water and are therefore confined in their distribution. In the Sudan Simuliids occupy a broad crescent of territory from the Ethiopian border to the Congo (Zaire)-Nile divide to around the upper part of the White Nile basin with northward extensions along the rivers hemmed in by increasing aridity. They are virtually absent in the great southern swamps and the White Nile downstream to Khartoum. Two of the species require attention – *Simulium damnosum* is a vector of *Onchocercus volvulus* in the south, though much less in its northern expanse, and *S. griseicolle* which appears in mass emergences in the northern stretch of the Main Nile from Debba downstream. In general the seasonal appearance of Simuliids depends on the breeding season, which in the south is during the rains from May to October, whereas in the rainless north it is from December to April with mass emergence before the flood of silt-laden water from the Blue Nile comes down from June to October. The extraordinary phenomenon of mass appearance has been commented on by every traveller who came from Egypt via the Nile route during the time of outbreaks. The following accounts may be quoted, separated by long intervals to show the permanence of the pest. An unnamed observer (Hill 1948 transl.) records on the 26th December 1824: 'We were tormented by small flies which infest the country for three months of the year' with dire effects on the riverain population. A few years later De Cadalvène and companion tell us about the 'innumerable swarms of a small fly . . . (at Argo Island) . . . which surround man and animals, getting into their eyes and ears causing painful stings'. In 1852 B. Taylor (1854) notes at Debba: 'I first made acquaintance with a terrible pest . . . a small black fly as venomous as the mosquitoes' . . . At Old Dongola further north 'I was so beset with the black gnats that I could not sleep . . .'. In 1906 Balfour during the 'stocktaking' of diseases in the newly re-occupied Sudan mentions the distribution of *Simulium damnosum* at Abu Hamed and further north, and that of *S. griseicolle* in the Dongola province from January to April, not spreading further away from the river than 'half a mile'. Fifty years later in 1956 the plague of Simulium remained unchanged (Rzóska 1964).

S. damnosum swarms and breeds in the Nile at Abu Hamed in the Nubian desert from 'October to June with a peak in February . . . causing intense skin irritations and suffering to people and animals' according to investigations in 1971 (Gassouma WHO/VBC/72.407); only little onchocerciasis is transmitted in this area. An outstanding biological problem remains unexplained – the pause of 8 or 9 months in which especially *S. griseicolle* disappears completely. In the south there is a dry season to account for little breeding but in the north with no rains and the river as sole habitat with a rigorous flood regime there must be some form of estivation so far not found.

Chironomidae

This is a group of great significance for the hydrobiologist because chironomids form a large percentage of the benthos of any waterbody and some species have been used as indicators for lake 'types' in the temperate zone.

Our taxonomic knowledge about chironomids in the Sudan stems mainly from Kieffer (1921, 1922, 1923 and 1924) and papers by Freeman (from 1954 onwards); Kieffer alone described 68 species of adults from the White Nile but some of his identifications are uncertain. A number of species were identified in the Blue Nile at Khartoum (see chapter 25).

In two places in the Sudan mass emergences have been observed – along the Blue Nile, especially at Khartoum, and 1,500 km. downstream at Wadi Halfa. Lewis has given detailed accounts of these in three papers (1954, 1956a, 1957). (Rzóska (1964) has summarised some of the results of hydrobiological interest. In cooperation with the government analyst at Khartoum, light traps were used to assess the number of chironomids emerging. The traps were operated in 1950/51 and 1951/52 from 18.30 to 20.00 hours every evening during the outbreak, which usually starts by the end of December and lasts until April. During the $1\frac{1}{2}$ hours of light trap operation up to one million insects were caught; 18000 of the small midges were dried and weighed 1g; on the basis of this unit numbers were calculated. Daily fluctuations in numbers seemed to be correlated with wind speed but the multi-species composition of the swarms did not allow for a more specific analysis. All seem to emerge from the Blue Nile and penetrate up to 300 m. into the townfront and die within a few hours, causing severe allergic symptions, especially asthma (Lewis *et al.* 1954).

The mass emergence of these and other insects means a loss to the benthos and the general ecology of the river as testified by the analytical work of Grindley (see chapter 25).

Searching for the causes of the outbreaks Lewis (1956a) traced some connections to the effect of dams both in the Blue Nile at Sennar erected in 1925, and at Wadi Halfa after the second heightening of the Aswan Dam in 1933. The increased reservoir was filled in 1935 and extended to a point near Wadi Halfa. Outbreaks noticed from 1938 onwards were so bad that in 1948 the transfer of the town of Wadi Halfa into the desert nearby was contemplated. Control measures at Khartoum undertaken in 1955 with Gammexane and in 1956, 1957–59 with DDT emulsion did not give conclusive results; in 1956 death of fish was recorded after DDT. In 1963 the WHO sent a specialist to examine the situation to suggest remedies including biological control (Wülker WHO/EBL/11-Vector Control/42/1963).

Other waterborne insect vectors

Mosquitoes are numerous with 156 species recorded by Lewis (1956b) and a number form part of the river fauna; 63 species living in the swamps of the Upper Nile, include vectors of malaria and yellow fever.

Only eight species of mosquitoes were found in 1943 at Wadi Halfa (Lewis 1944) of mixed palearctic and 'Ethiopian' composition. This number of species increases in the central and southern Sudan; as mentioned most of these are 'Ethiopian'.

Tsetse is distributed in a belt along the southern frontiers of the Sudan – the distribution is associated with woodlands along streams. The two main species are *Glossina morsitans* and *G. palpalis* but three more species may be involved as disease vectors. Sleeping sickness was once widespread but is now reduced, though it still occurs in sporadic outbreaks like that reported at Maridi in 1945. Of major importance is cattle trypanosomiasis affecting also horses, which precludes the keeping of these animals with serious repercussions on the protein supply of some tribes, e.g. the Azande on the Congo (Zaire) border. In 1948 70% of the 180000 people there had to be resettled because of Tsetse according to Ferguson (1949, Empire Cotton Growing Review 26: 109–121).

Phlebotominae, sandflies, are common in the Sudan. Though they breed in the soil, the distribution of some species is affected by the water table and proximity of rivers. Sandflies include vectors of oriental sore, Kala-azar and sandfly fever.

Ceratopogonidae, biting midges, a sub-family of the chironomids, suck blood and can be a pest locally; 30 species have been recorded by Macfie in 1947 (Proc. Roy. Ent. Soc. b 16: 69–78) and Austin in 1943 (Bull. Ent. Res. 3: 99–111).

Of other biting flies the Hippoboscidae, Stomoxys may be mentioned for their impact on human and domestic stock.

In spite of the great efforts of the World Health Organisation (WHO) serious outbreaks of tropical diseases still occur. Along the northern Nile a spread of malaria into Egypt occurred in 1942; Ethiopia had a recent epidemic of yellow fever; in Uganda the great outbreak of sleeping sickness in the early 20th century is still remembered.

EDITOR. The account given here is based on the multitude of reports, papers, field observations by a great number of entomologists, veterinary and medical officers, and administrators since the beginning of the century. A considerable number of these investigations have only been recorded in government files. This material has been sifted by D. J. Lewis, who was medical entomologist in the Sudan for 20 years; he has added his own research results in a series of papers summarising these efforts. These papers are also a rich source of general ecological and historical information on the Sudan.

REFERENCES

There is an enormous literature on insects of medical importance for the Nile basin. Only some are included; the papers by D. J. Lewis contain a rich bibliography.

Baker, Sir S. W. 1871. The Nile tributaries of Abyssinia . . . 4th edition London.
De Cadalvène, E. & de Breuvery, J. 1836. L'Egypte et la Turquie de 1829 à 1836, vol. 1 and 2. Egypte et Nubie. Paris.
Gleichen, C. 1905 (ed.). The Anglo-Egyptian Sudan. A compendium prepared by officers of the Sudan Government. vol. 1: Geographical description and History. vol. 2: Routes of reconnaissance. Harrison & Sons, London.
Hill, R. L. 1948. An unpublished itinerary to Kordofan 1824–1825. (Translation). Sudan Notes and Records 29: 58–70.
Lewis, D. J. 1944. Observations on Anopheles gambiae and other mosquitoes at Wadi Halfa. Trans. Roy. Soc. Trop. Med. and Hygiene 38: 216–229.
Lewis, D. J. 1952. Early travellers accounts of the surret flies (Tabanidae) in the A.-E. Sudan. Sudan Notes and Records 33: 276–298.
Lewis, D. J. 1953a. Simulium damnosum and its relation to onchocerciasis in the A. E. Sudan. Bull. Ent. Res. 43: 597–644.
Lewis, D. J. 1953b. Tabanidae in the A. E. Sudan. Bull. Ent. Res. 44: 175–216.
Lewis, D. J. 1953c. Simuliidae in the A. E. Sudan. Rev. Zool. Bot. Afric. 48: 269–286.
Lewis, D. J. 1956a. Chironomidae as pest in the northern Sudan. Acta Tropica 13: 142–158.
Lewis, D. J. 1956b. Some mosquitoes of the Sudan. Bull. Ent. Res. 47: 723–735.
Lewis, D. J. 1957. Observations on Chironomids at Khartoum. Bull. Ent. Res. 48: 155–184.
Lewis, D. J., Henry, A. J. & Grindley, D. N. 1954. Daily changes in the numbers of Chironomid midges at Khartoum. Proc. Roy. Soc. Ent. Lond. (A) 12: 124–128.
MacMichael, Sir H. 1934. The Anglo-Egyptian Sudan. Faber & Faber, London.
MacMichael, Sir H. 1922. A history of the Arabs in the Sudan and some account of the peoples who preceeded them . . . Cambridge Univ. Press.

Rzóska, J. 1964. Mass outbreaks of insects in the Sudanese Nile basin. Verh. Intern. Ver. Limnol. 15: 194–200.
Taylor, G. 1854. Life and landscape from Egypt to the Negro-kingdom of the White Nile . . . Sampson, Low & co., London.

(The taxonomic papers by Kieffer, Freeman and others are not listed here, this bibliography is contained largely in Lewis 1957).

24. ZOOPLANKTON OF THE NILE SYSTEM

by

J. RZÓSKA

The zooplankton of the whole riversystem is treated in one chapter, because the two main rivers bind the various lakes and other features

Table 1. Planktonic Crustacea of the Nile System.

Species	Lake Victoria	L. Kioga	L. Albert	white Nile	Blue Nile	L. Tana	Egypt. Nile
Tropocyclops prasinus Fisch.	+			+			
Tr. confinis Kiefer	+						
Mesocyclops leuckarti Claus at least two forms	+		+	+	+	+	+
Thermocyclops hyalinus consimilis	+						+
Th. neglectus and varieties (Sars, Kiefer)	+			+	+		
Th. infrequens Kiefer	?						
Th. schuurmanae (Kiefer)	+		+				
Th. schmeili Poppe & Mrazek	+						
Th. emini Mrazek	+			?			+
Thermo-diaptomus galebi Barrois				+	+	+	+
Th.-diapt. galeboides Sars	+						
Th.-diapt. stuhlmanni Mrazek	+						
Tropodiaptomus syngenes Kiefer				+	+		+
Trop.-diapt. kraepelini Poppe & Mrazek				+	+		
Trop.-diapt. processifer Kiefer				+	+		
Trop.-diapt. orientalis Brady Sars				+	+		
Diaphanosoma excisum Sars	+	+ and D. sarsi	+	+	+	+	+
Ceriodaphnia cornuta Sars	+	+	+	+	+	+	+
Cer. dubia de Guerne & Richard	+			+	+	+	+
Cer. reticulata (Jurine)			+			?	
(*Moina dubia* de Guerne & Richard)	+			+	+	+	+
= *M. micrura* Kurz	+	+					
Daphnia longispina Leydig	+					+	+
D. lumholtzi Sars	+		+	+	+	+	+
D. barbata (Weltner)	+			+	+		+
Chydorus sphaericus O.F.M.	+						
Bosmina longirostris O.F.M.	+	+		+	+	+	+

into some entity. This is not so in other biological features; fishes and part of the benthos are differentiated in this great variety of waters. Plankton and other suspended material are carried by the current of rivers and therefore show some uniformity.

Rivers carry suspension composed of non-biotic elements, adventitious organisms from the shore and the bottom, and of truly planktonic forms. In a long river these elements separate under suitable conditions. In the Nile such a true plankton exists in varying densities from the headwaters to the ultimate end of the river system.

Specific remarks on main sites

The following remarks are based on material collected with different intensity along the Nile. The most intensively studied area is in the Sudan stretch of both the White and Blue Nile; hundreds of samples, both net- and quantitative, have been studied. The material of the other regions will be mentioned in a survey of the sites. The composition of the Crustacean zooplankton is presented in Table 1.

Some remarks to this species list are necessary. The identification of Entomostraca in the tropics suffers from the lack of finer taxonomic, morphological and physiological studies. Where applied, as in some groups of Cyclopids, especially the genus *Thermocyclops* and others, considerable diversion from similar temperate-zone have been found; but morphology and taxonomy are at present unfashionable with detrimental effect for ecological studies.

With these reservations the zoogeographical composition of the zooplankton is discussed. The following Cladocera are circum-tropical: *Diaphanosoma excisum, Ceriodaphnia dubia, C. cornuta* (comprising the *rigaudi* form), *Moina dubia = U. micrera*; a large host of shore Cladocera belongs to this group. A further group is regarded at present as cosmopolitan: *Bosmina longirostris*, forms of *Simosa, Chydorus sphaericus* and many of the small shore Chydoridae. Zoogeographically confined is *Daphnia barbata* which is African and *Daphnia lumholtzi* which is North-African and Middle-Eastern. – The Cyclopids contain some cosmopolitan forms, the Diaptomids are specific; Lindberg and Kiefer have contributed greatly to a better understanding of finer differentiations and speciation of some groups in Africa.

Of the species found in Lake Victoria the following have not been found in the Nile of the Sudan: *Chydorus sphaericus, Daphnia longispina* (but found in Lake Tana and Egyptian waters), *Tropocyclops confinis, Thermocyclops emini*, and *Th. schuurmannae, Thermodioptomus galeboides, Thermodioptomus stuhlmani*. Green (1971) mentions a number of species (forms) from L. Victoria, Kioga and Albert which have also not been seen in the river plankton of the Sudan. On the other hand the 4 species of *Tropodiaptomus* mentioned in the list of Nile plankton have not been recorded

334

from the headwaters; in mass samples of plankton species can be easily overlooked. – Some algal species of L. Victoria have also not been found in the Sudan Nile. A list of the Rotifers will not be given; over 60 species have been recorded from the Nile system so far but in view of their wide distribution all over the world and continuous additions such list is not important. Berzins (1955) has found some distribution differences in *Keratella valga*, mainly palaearctic, and the more tropical forms *K. tropica*, *K. lenzi* and *procurva*. Green (1972), comparing material from a number of global sites and using the Sorensen index, found latitudinal variations in 8 genera of Rotifers with some forms preferring warm other cold climates. Both these studies incorporated Nile material. The Rotifers of Egyptian waters have been studied by Klimowiez (1961, 1962).

Some additional plankton components should be mentioned: *Limnocnida tanganyikae*, the fresh water medusa, has been recorded from Lake Victoria but not other parts of the Nile system (Pitman 1965); *Caridina nilotica* seems to form a pelagic appearance there and in L. Albert; there are pelagic fishes in L. Victoria and L. Albert; plankton stages of Chaoborids are widespread throughout (see ch. 25). – Adventitious forms in the plankton are locally numerous depending upon current conditions; Turbellaria, recorded often in the deeper layers of rivers, should be mentioned specifically.

Lake Victoria

Thirty years after the discovery of this lake the first plankton net was dipped into it by Emin Pasha and F. Stuhlmann in 1888. Then after some interval many, mainly German, expeditions brought back samples. The results have been surveyed by Rzóska (1957), and the list of main species was largely established. But it is a remarkable fact that no intensive studies on the zooplankton of Lake Victoria have ever be carried out and we know nothing about distribution, seasonal changes and production. In the last 45 years only three papers have dealt with the zooplankton of Lake Victoria. Worthington (1931) established a diurnal vertical migration at one open lake station and gave some quantitative figures on the basis of net samples; Green (1971) examined numerical appearances at several inshore stations; Rzóska (1957) examined samples from open lake and inshore stations to find out about the distribution of zooplankton. All these have only fragmentary significance, the older contributions have only limited significance.

Lake Kioga

Green (1971) examined zooplankton from 7 stations in one of the northern arms of this dendritic lakes and found *Diaphanosoma Sarsi* besides some usual components; no *Daphnia* was recorded in any sample.

He gives numbers of Crustacea under 1 m² on the basis of net hauls. Rotifers of this and other headwater lakes have been discussed by him also (Green 1967b).

Lake Albert

Here again Green has contributed some recent studies on the Cladocera and Rotifera of the 'lake sources' of the White Nile (1967a, b). In L. Albert 8 stations with a number of transects allowed him not only to establish a species list (with no remarkable deviations from the usual set) but to discover finer trends of distribution of two forms of *Daphnia lumholtzi* (1967a). A helmeted form lives in inshore areas, a helmless form ('monacha') in off-shore areas. Green sees some advantages against predation by the smaller less conspicuous helmeted from. Here again numbers of Crustacea are given per units of watermass; the Sørensen index is applied to zooplankton of four other lakes to show a greater diversity in waters with a discernible flow.

His most northerly station is on the Albert Nile at Pakwach where a good plankton association flowed down the Nile. Spot samples taken by Rzóska at the Lake Victoria outflow, at Rhinocamp and at Nimule in 1949 (unpublished) showed the persistence of a plankton, though with increasing river conditions the admixture of detritus and adventitious forms was noted. After the descent through the Nimule gorge (see Fig. 32) there was only a scanty suspension visible but with a few true plankton forms.

Bahr el Gebel

In this stretch in the Sudan plains we have the usual picture of a scanty river plankton, which builds up in the swamps with its numerous interspersed standing waters. In these, with currents reduced or absent, sorting out occurs of the suspension and pure plankton associations develop in profusion. Together with the Ghazal plankton and that from the Sobat, the White Nile is well supplied to form the great development of plankton in the Gebel Sulia basin and the final stretch of the White Nile (see ch. 12, 13).

In the Blue Nile, Lake Tana has a pure zooplankton as described in ch. 14, the survival of rudiments in the Blue Nile Gorge is extraordinary but a fact, in view of the numerical developments downstream (ch. 16).

The joint Nile from Khartoum to the Egyptian border has only been investigated by few spot samples; but enough is known to explain the persistence of plankton and its occasional strong development in the north where the river is hemmed in by rocky gorge-like shores in the cataracts, slowing down currents. Now with the great dam basin of Aswan reaching deeply into the Sudan lake conditions prevail with all the consequences entailed. Below Aswan previous conditions of a considerable zooplankton

Elster & Vollenweider 1961) are now greatly reinforced by the great development of plankton in the High Dam basin (see ch. 19). It seems that none of this biological production enters the sea anymore and is distributed with irrigation water (Klimowiez 1961). Most of the species from the African hinterland waters appear in Egypt; some additional species have been noted (Salah & Tames 1970).

General remarks

We can now sum up some general remarks on the zooplankton phenomenon in the Nile system as far as they are known from the results in the different sectors. The water regime is the main factor influencing the development of plankton; temperatures with their narrow amplitudes in the Nile basin are not decisive; for phytoplankton, of course, nutrient supply is a major factor.

Generally, wherever current is slowed down, be it by a flat slope, meandering, obstacles both natural, like narrow cataracts, or man-made dams, a true plankton association develops. This process has been followed in detail in the Ghazal river as natural sequence, and in the Gebel Aulia basin as man-made sequence (Fig. 82, 83). The carrying capacity of discharge and current drops, the water clears and the process of biological production starts. At present the Nile system is divided in stretches of river conditions and four large dam basins with flowing lake conditions; in addition there are the headwater lakes and the numerous standing waters on the White Nile.

The origin of river plankton was once an issue of controversy; it is not anymore. The supply to this enormous film of water moving downstream over thousands of km cannot be pinpointed in detail; its inoculum is derived obviously from the lakes and standing waters where plankton is permanent to rivers where it persists. Puzzling is the reconstitution of the plankton in the Blue Nile, largely but not completely obliterated by the annual flood. Here river inlets harbour remnants of plankton which seem to be sufficient to repopulate the river after the flood; Asim Moghrabi (unpublished) thinks that 'diapause' stages are important, formed by the constituents of the Blue Nile plankton. This maybe so in some species e.g. Cladocera with ephyppia and some resting stages in Copepods. But the main factor must be sought in quick reproduction processess from surviving specimens and possibly resting eggs.

Examination of many hundreds of samples has revealed the presence of ephyppia in all species of Cladocera of the Nile Plankton; these appear at various times of the year and, not like in temperate waters, under climatic stress. Nor are epphyppia formed specifically in population stresses e.g. during 'mass appearances'. These occur in the Nile but in moderate form. Males have been noted also in all Nile Cladocera but no clear trend can be deduced. When Nile river water is pumped into fishponds sexual

337

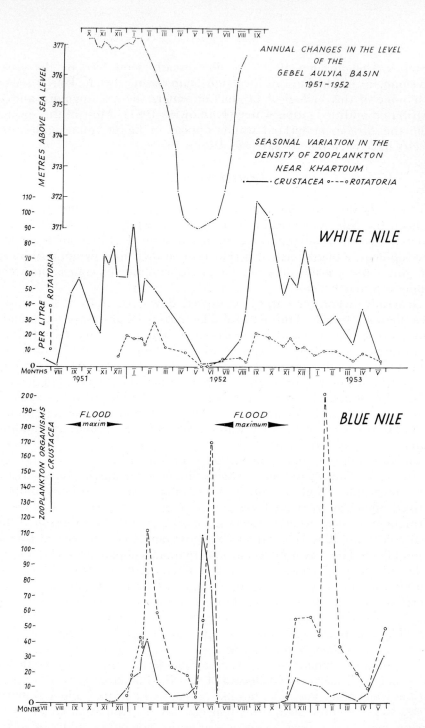

Fig. 82. Seasonal plankton development in the White and Blue Nile at Khartoum in 1951–1952. Note influence of the Gebel Aulia dam regime and of the flood of the Blue Nile; from Rzóska, Brook & Prowse 1955.

338

processess with males and ephyppia increase greatly together with population densities. Copepods are bisexual and the relation of the sexes sometimes shows imbalances (Rzóska, unpublished).

In standing waters of the Nile the continued presence of plankton does not require comments. In running water a limited time is available for the floating community. A great number of calculations have been made to find out the time of flow in some sectors. These are of limited value because based mainly on midstream currents calculated from discharges (Nile Basin volumes) and travel of discharge fluctuations. Our own observations on velocities of water flow showed a great unevenness both horizontal and vertical (Rzóska 1974) besides the season differences of low and high water conditions. With these reservations some select figures on time of travel may be mentioned.

White Nile: Mongalla Lake No, 750 km, travel time of level fluctuations 35 days; by midstream currents 9–13 days; Kosti-Gebel Aulia dam when closed, 272 km, by currents 22 days; Retention time in Gebel Aulia basin about 40 days (theoretical calculation); Ghazal river, 200 km, currents between 0.3–0.018 m/sec, 17–40 days.

Blue Nile: Sennar to Khartoum, 357 km, at low river more than 40 days; at flood only $2\frac{1}{2}$ days.

Incidentally the brown colour of the flood is similar to the colour indicators used in research of water movements. This brown flood water arrived at Khartoum in 1967 about the 15 June and at the Wadi Halfa reach of Lake Nubia on the 27 June; these 12 days tally with a 10 days calculated from current velocities. These times are now greatly delayed. But all these values are relative and there are quoted here only to show that enough time is available in the sectors of the Nile to allow plankton to develop.

That zooplankton can succeed to populate the river in the time available is clear from its presence. But to find the reproductive rates, enabling this, seems to be impossible in a environment which passes away continuously. Incidental observations on the speed of the development have been made in pools left by the falling river (Talling & Rzóska 1967); 12 days after their isolation they contained fully breeding specimens of three most important plankton constituents *(Thermocyclops, Moina, Diaphanosoma)* at least in their second generation. Under extreme conditions of rainpools *Moina* took only 48 hours from hatching to egg-gearing (Rzóska 1958).

From the remarks made above some general principles emerge of plankton distribution along the Nile.

Distribution of zooplankton in the Nile system

Many quantitative samples have been collected along the Nile system and some have been used for illustrating seasonal changes in densities and for biomass assessment e.g. by Monakov (1969). These I find risky for reasons which are discussed below in detail but which can be summarised briefly as uneveness of distribution. In flowing water this is understandable; their peculiar time/space relationship is admirably summed up by a none-biologist: Leonardo da Vinci (Codice Trivulziano fol. 34 and 'Selections from the notebooks of L. da V.', World Classics Oxford Univ. Press 1953). He says: 'In rivers the water you touch is the last of what has passed and the first of that which comes; so with time present'.

In other words a sampler or net dipped at intervals collects a different water mass, and more different the longer the interval. With this in mind let us group the different modes of distribution which have been observed in the river Nile.

1. A seasonal succession each year caused by hydrological factors, the flood in the Blue Nile and the change from the 'flowing lake' to river condition in the White Nile. The papers published on the seasonal development of the plankton contain data amply illustrating this phenomenon (Rzóska et al. 1955, see Fig. 82, Prowse & Talling 1958, Talling & Rzóska 1967, Hammerton 1972 and ch. 16).

2. A spatial built up over long distances caused by natural factors, stowing effect of other waters on slow currents as in the Ghazal river, or narrowing of river bed as in the Dongola reach of the northern Nile. But artificial barriers, dams, are the most striking example and this happens in the Nile in the Gebel Aulia basin (see Fig. 83) and at Roseires and Sennar (see ch. 16). This is the most predictable phenomenon in the biology of rivers.

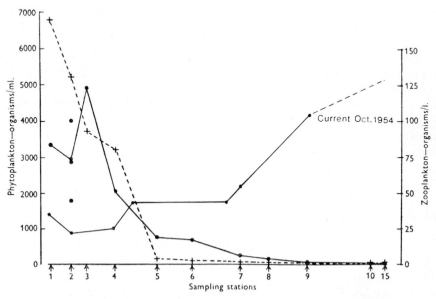

Average density of phytoplankton (+ – – +) and zooplankton (●—●).

340

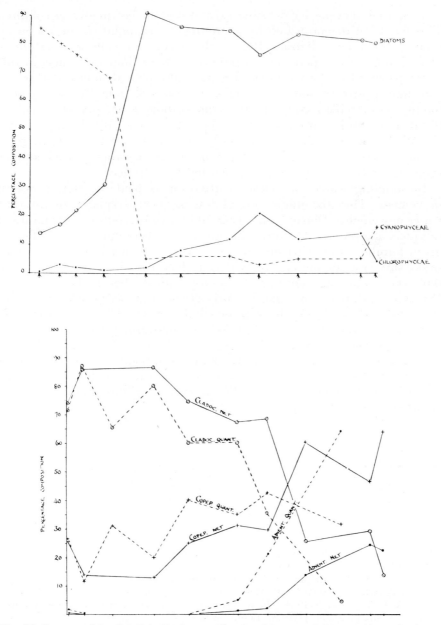

Fig. 83. Impact of the Gebel Aulia dam on the development of phyto- and zooplankton, 13–18 December 1951. Upper graph represents total numbers per ml or litre, the two lower graphs changes in composition; the lowest graph compares also the results of net samples and quantitative samples. Sampling stations are: 1. Khartoum, 2. Gebel Aulia (44 km), 4. Geteina (94 km), 7. Kawa (224 km), 9. Kosti (319 km), 10. Gebelein (394 km), 15. Keri kera (430 km from Khartoum). After Brook & Rzóska 1954 and unpublished data.

3. Spatial uneveness of densities exist also in the flowing water column. This has been demonstrated by sampling at one point in succession of minutes or hours (Rzóska 1968); the results of collecting 20 one-litre samples at one point are unpublished. Both show uneveness in the water passing by. Spatial differences have also been demonstrated when travelling upstream and collecting at intervals of time; Rzóska (1974) found considerable changes in the composition of net plankton over a distance of 160 km, traversed in 11 days. Horizontal distribution across the White and Blue Nile proved very uneven in two transects in 1952 (unpublished). The vertical differentiation in densities is well known though in rivers this is largely obliterated by turbulence.

In standing waters not enough attention is paid to internal water movements. Here the unidirectional flow of rivers is replaced by internal water movements. Diurnal migrations are based on counts of net hauls or samplers and one would expect that the counted columns would reflect a given number of animals moving up and down. In fact the number of organisms in a column can be widely different at the same time of the day when sampling continues for more than 24 hours. The results in Worthington's paper on Lake Victoria (1931, a pioneer work for Africa) show differences up to 10 times in the whole series and 5 times at the same time on successive days. Calculations on some European migration series show the same discrepancies.

In conclusion: zooplankton in rivers floats in bouts or 'clouds'; it thins out and thickens when current slows down; it renews itself after obliteration with astonishing regularity. The uneven distribution applies also to standing waters of the Nile system.

REFERENCES

Abu Gideiri, Y. B. 1969. Observations on the zooplankton distribution at the 'Mogren', Khartoum. Rev. Zool. Bot. Afric. 79: 1–2.
Berzins, B. 1955. Taxonomie und Verbreitung von *Keratella valga* und verwandten Formen. Ark. f. Zoologi 2: 549–559.
Brook, A. J. & Rzóska, J. 1954. The influence of the Gebel Aulia Dam on the development of Nile plankton. J. Anim. Ecol. 23: 101–115.
Elster, H. J. & Vollenweider, R. 1961. Beiträge zur Limnologie Aegyptens. Arch. Hydrob. 57: 241–343.
Green, J. 1967a. The distribution and variation of *Daphnia lumholtzi* (Crustacea: Cladocera) in relation to fish predation in Lake Albert, Uganda. J. Zool. Soc. Lond. 151: 181–197.
Green, J. 1967b. Associations of Rotifera in the zooplankton of the lake sources of the White Nile. J. Zool. Soc. Lond. 151: 343–378.

Green, J. 1971. Associations of Cladocera in the zooplankton of the lake sources of the White Nile. J. Zool. Soc. Lond. 165: 373–414.

Green, J. 1972. Latitudinal variation in associations of planktonic Rotifera. J. Zool. Soc. Lond. 167: 31–39.

Hammerton, D. 1972. The river Nile – A case history. In: River Ecol. and Man. Acad. Press, New York – London, 171–214.

Lindberg, K. 1956. Cyclopides de L'Ouganda. Kungl. Phys. Sällsk. Fordhdlg. Lund 26: 1–14, 25–38.

Kiefer, F. 1952. Copepoda Calanoida und Cyclopida. Explor. Parc Nation. Albert, Mission H. Damas, fasc. 21.

Klimowiez, H. 1961. Rotifers of the Nile canals in the Cairo environs. Polsk. Arch. Hydrobiol. 9: 203–221.

Klimowiez, H. 1961. Differentiations of Rotifers in various zones of Nile near Cairo. Polsk. Arch. Hydrobiol. 9: 223–242.

Klimowiez, H. 1962. Rotifers of small water bodies of Cairo botanical gardens. Polsk. Arch. Hydrobiol. 10: 241–270.

Moghrabi, Asim Ibr. 1972. The zooplankton of the Blue Nile. Ph.D. Thesis, Fac. Sc. Univ. Khartoum (unpubl.).

Moghrabi, Asim Ibr. (unpubl.). A study in diapause of zooplankton in . . . the Blue Nile. Hydrob. Res. Unit, Univ. Khartoum.

Monakov, A. V. 1969. The zooplankton and the zoobenthos of the White Nile and adjoining waters in the Rep. of the Sudan. Hydrobiologia 33: 161–185.

Pitman, C. R. S. 1965. The African freshwater medusa *Limnocnida tanganyikae* Günther, and a general note on freshwater medusae. The Puku, Occas. Papers, Dept. Game and Fisher., Zambia, 3: 113–124.

Prowse, G. A. & Talling, J. F. 1958. The seasonal growth and succession of plankton algae in the White Nile. Limnol. and Oceanogr. 3: 223 238.

Rzóska, J., Brook, A. J. & Prowse, G. A. 1955. Seasonal plankton development in the White and Blue Nile near Khartoum. Verhdlg. Int. Ver. Limnol. 12: 327–334.

Rzóska, J. 1957. Notes on the Crustacean plankton of Lake Victoria. Proc. Linn. Soc. Lond. 168: 116–125.

Rzóska, J. 1961. Observations on tropical rainpools and general remarks on temporary waters. Hydrobiologia 17: 265–280.

Rzóska, J. 1968. Observations on zooplankton distribution in a tropical river dam-basin (Gebel Aulyia, White Nile, Sudan). J. Anim. Ecol. 37: 135–198.

Rzóska, J. 1974. The Upper Nile swamps, a tropical wetland study. Freshw. Biol. 4: 1–30.

Salah, M. & Tamas, G. 1970. General preliminary contribution to the plankton of Egypt. Bull. Inst. Oceanogr. Fisheries (Alexandria) 1: 305–337.

Stuhlmann, Fr. 1891. Beiträge zur Fauna zentral-Afrikanischer See. I. Südcreek des Victoria-Nyansa. Zool. Jhrb. 5: 924–926. (also, Vorläuf. Ber. 1888, Sitzber. Preuss, Ak. and Ibid. 1889).

Talling. J. F. & Rzóska, J. 1967. The development of plankton in relation to hydrological regime in the Blue Nile. J. Ecol. 55: 637–662.

Worthington, E. B. 1929. The life of Lake Albert and Kioga. Geogr. J. 74: 109–132.

Worthington, E. B. 1931. Vertical movements of freshwater macroplankton. Int. Rev. Hydrol. and Hydrobiol. 25: 394–436.

25. NOTES ON THE BENTHOS OF THE NILE SYSTEM

by

J. RZÓSKA

Whereas zooplankton in the Nile system moves with the flow of water connecting the various lakes and river stretches, the benthic fauna is mainly stationary and therefore will be locally differentiated, although drift may occur in some places. Two major habitats harbour benthic fauna, the shore and the deeper sediments. Many investigations in temperate-zone lakes have shown that both these habitats differ greatly although merging along the basins slope.

From the discovery of the East African lakes onwards many publications have appeared based on expeditions type of work; although these investigations inaugurated hydrobiological exploration in Africa they are mostly out dated as far as invertebrates are concerned. Modern work is concerned mainly with particular species or groups but no quantitative investigations of benthos have been carried out in the waters of the Nile system except by Monakov (1969) on the White Nile. But faunistic and ecological studies exist on some parts of the shore fauna e.g. in Lake Victoria and the upper Nile swamps. The present attempt must therefore deal with fragments of knowledge: it cannot give a coherent picture of the whole riversystem of over 6,000 km with a variety of waters, standing and running.

The shore fauna of the Nile system is rich and varied where there is a fringing vegetation as in Lake Victoria and along parts of the White Nile; it is poor in the Blue Nile river with partly steep shores and drastic changes of water level. The richness and variety of the shore fauna in the Upper Nile swamps has been described in ch. 12, and in more detail by Rzóska (1974). Almost every invertebrate group is represented in profusion; key animals are the Nile shrimps *Caridina nilotica*, the rarer *Palaemon niloticus*, the Conchostracan *Cyclestheria hislopi*, 26 genera of littoral Cladocera, a multitude of insects. Shoals of fish fry and smaller species of fish and other groups live here in a bewildering complexity of food relations.

This assembly, with some differences, populates the shores of Lake Victoria Kioga and Albert; in particular the littoral Cladocera found by Thomas (1965) in the fringes of Ugandan swamps live also in the 'Sudd' fringes.

In Lake Victoria extensive studies of mainly littoral insects have been conducted by the technique of light traps at the Jinja Station, near the outlet of the Victoria Nile. P. S. Corbet & A. Tjønneland have described

345

the results of their investigation from 1955 to 1966 in 22 and 10 papers respectively starting in 1955 with a joint paper.

These observations deal with emergences adults (or near adults) of species of Ephemeroptera, Trichoptera, Odonata, Chironomidae, Culicidae and other insects, whose larvae form a large part of the benthos. Times of emergence, daily and lunar rhythms, sequence of generations, food relations and composition of mass samples collected are treated. All this work provides only indirect information on the benthos actually populating the lake. Nor is it possible to compile a reliable list of species forming the shore fauna. Even if all publications would be assembled the list would be still un-representative. Taxonomic determination, necessary to support ecological studies, is far from achieving a truly representative list of the fauna. New species are found by specialists continuously; e.g. material sent to D. E. Kimmins (of the Brit. Mus. of Nat. Hist.) and published in from 1953 to 1960 contained 18 new species of Trichoptera out of 39 found and 14 new species out of 40 present of Ephemeroptera. The same is happening in other insect groups associated with water, especially in the difficult Diptera. The larvae of the Chironomidae and Culicidae are an important part of the benthos and in the well known temperate lakes have been used as indicators of lake 'types'. But this cannot be extended to tropical waters with any certainty at present. About a dozen of different Chironomid species have been studied but mostly identification denotes only a generic or group name. Their quantitative distribution in the bottom sediments seems to have been treated only by MacDonald (1956). *Chaoborus* larvae (Culicidae) showed densities of 2000 to 2500 per m², Chironomid larvae up to 1000; similar densities persist still at 30 m depth in Lake Victoria.

Although mass appearances of aquatic bred insects are well known in many parts of the world but in Lake Victoria they supply information about the largely unexplored bottom fauna. 'Lake fly' swarms have been observed for many years, Macdonald (1953) has examined their composition; he also found that these swarms over the lake occur with some periodicity. Besides several species of *Chaoborus*, a number of Chironomids take part in the swarms. The ecology of individual species in these mass emergences has been the subject of the papers by MacDonald, Corbet, Tjønneland and others. These conspicuous swarms (Fig. 84) have also an impact on man. With the steadily increasing human population around the lake these swarms have become a nuisance; Tjønneland quotes a technical report on the construction of the Owen Falls dam, where non biting insects caused considerable interference with the construction work. The appearance of *Simulium* was so severe that large scale eradication measures had to be made.

But the value of these lake-dwelling insects lies, of course, in their supply as food organisms for fishes and other members of the community. MacDonald (1956) has investigated the importance of *Chaoborus* larvae for

346

Fig. 84. 'Lake flies', swarms emerging from Lake Victoria; Photo P. H. Greenwood.

five species of the Mormyridae; the number of larvae in the sediments, with regular peaks of generations, is reflected in the food recorded in the stomachs of these fishes. The production is great, Corbet (1964) wrote: 'The rapid growth and replacement of larval populations in tropical lakes must be of greatest importance in effecting a rapid turnover of nutrient substances and in maintaining a large and constant supply of these in the form of available animal protein'.

There are no extensive studies on the Oligochaeta of Lake Victoria but see some remarks in ch. 6d. by S. Ghabour; as already mentioned there has never been a special study of the deeper bottom fauna of this great lake. For the Crustacea two recent contributions exist; it has already been mentioned that species of the Atyidae are key animals in the shore benthis community and G. Fryer (1960) has written a beautiful study of the feeding (filtering) mechanisms of two species of *Caridina*. Crabs of the Nile system are treated in a special contribution on later pages. The important molluscs have been investigated by Mandahl-Barth (1954) and a summary of his work is presented here. He lists 126 species of molluscs, 86 gastropods and 40 bivalves, in Uganda and adjacent territories, of these 65 species and subsp. live in Lake Victoria. The distribution is mainly littoral and in shallow waters; only 4 species have been found to a depth of 70 m; these are *Melanoides tuberculata, Bellamya unicolor, Sphaerium stuhlmanni* and *Corbicula africana*; *Pisidium fistulosum* lives only in the depths of the Lake. Lake Victoria shows a splitting of species which is not repeated with the same intensity in the two lakes connected with it.

347

Table of speciation – Mollusca, Gastropoda.

| Family | species or genus | 'Forms' appearing in | | |
		Lake Victoria	L. Kioga	L. Albert
Viviparidae	*Bellamya unicolor*	13	1	1
Ampullariidae	*Pila ovata*	4	2	2
	Lanistes carinatus	0	1	1
Bytiniidae	*Gabbia*	1	0	4
Thieridae	*Melanoides tuberculata*	2	1	2
	Cleopatra	4	0	1
Limnaeidae	*Limnaea*	3	2	1
Planorbidae	*Biomphalaria*	6	2	4
Pulmonata	*Anisus*	1	1	1
	Gyraulus	7	1	1
	Lentorbis	1	1	1
	Segmentorbis	2	1	1
Bulinidae	*Bulinus (Pyrophysa Physiopsis)*	10	5	3
Ancylidae	*Burnupia*	2	0	0
		56	18	23

Species and subsp. of *Bulinus* and *Biomphalaria* are vectors either for *Schistosoma haematobium* or *S. mansoni*. Molluscs contribute greatly to the food-webb, there are 8 mollusc eating fishes in the area discussed (see ch. 7, Fig. 13); for Schistosoma in general see ch. 22. – The splitting into races in molluscs is regarded by Mandahl-Barth as outcome of level changes and separation of the lakes into basins; Lake Albert has been richer in the past as seen by remnants in the subfossil Kaiso beds.

Special mention must be made of the occurrence and remarkable developmental history of the bivalve *Mutela bourguignati* (Ancey) Bourguignat; this species lives in rocky sites of Lake Victoria and the Victoria Nile. Material from these sites allowed G. Fryer (1961) to unravel the parasitic phase which the larva undergoes on the cyprinid fish *Barbus altianalis radcliffi* until the metamorphosis into a juvenile small mussel. For the first time the whole life history has been followed in this study.

Benthos of other parts of the Nile basin

The only quantitative investigations of the benthos of the Nile system, except some few data from Lake Victoria mentioned and some from Egyptian inland waters (ch. 20), are those by Monakov (1969). He took about 300 samples with a Petersen grab and calculated the biomass per m^2 from the number and the dry weight of the animals. The paper gives some valuable information on local distributions in standing and running waters of 15 sites in the White Nile from Bor to the Gebel Aulia dam.

Faunal lists are attached, which in some groups need taxonomic revision. One of the best investigated localities is Lake No, where a map illustrates the distribution of sediments, brown and grey, and the corresponding populations of benthic animals. Monakov expresses in weight the great difference between the rich shore fauna amidst vegetation and the much poorer bottom faunas; the difference may be as much as 10 to 100 times in weight. He notices the increase of benthos of the river south of the Gebel Aulia dam where the current slackens and the sediment falls out. (Ed. Incidentally this is the region where fishing camps were set up with considerable success). Further, a general increase of the bottom fauna towards the middle part of the Gebel Aulia basin is demonstrated in histographs. But although this is a very valuable study its absolute values of biomass are only valid for the time of sample taking; this is certainly so with his biomass figures for zooplankton; in zoobenthos Monakov himself notices considerable changes of biomass in two successive visits and suggests especially emergences of Chironomids as causes of differences.

That this so has been amply demonstrated in the Blue Nile where mass emergencies of Chironomids were studied because of ill-effects on humans. The general aspects are discussed in ch. 23, here some facts, relating to their significance as benthos, must be briefly summarised.

Lewis (1954, 1957) gave the first scientific observations on the composition, seasonal appearance, distribution and origin of these swarms, noticeable especially at Khartoum. With the help of Freeman the presence of about 25 species or forms of adult midges (Rzóska 1964) was established at Khartoum. Amongst these *Tanytarsus lewisi* was a prominent component. In view of the seriousness of the situation, including asthma outbreaks, control measures were applied such as Gammexane powder and later DDT. Several series of bottom samples were taken by Rzóska (unpublished) to find out the distribution of larvae; some of these gave densities up to 90,000 per m². As the measures taken were not effective, a WHO expert, Dr. Wülker, reexamined the situation later (1963). He investigated the distribution of larvae on the river bottom, added a number of species to the list and recommended further measures, including biological control. – The so far unexplained phenomenon is the disappearance of larvae during a large part of the year, with only few specimens present, and the mass appearance of them and the adults from November to March. No doubt exists as to a diapause phase, sofar not found the adults each year from November until March in successive waves of species; only a diapause phase can explain this, but has not been found sofar. Similar outbreaks of other insects in the northern Sudan are described in ch. 23.

It is obvious that such emergences cause considerable changes in the benthos seasonally and are a considerable loss of biomass from the aquatic habitat. The changes in emergences were examined in a series of experiments at Khartoum (Lewis et al. 1954). The Chironomids emerge from the Blue Nile shortly after sunset and are actively around for about 1½ hours before they die. With the help of a light trap, collections were made and analysed as to their body composition (Grindley 1952). Dead insects were oven dried at 50 °C and weighed; 18000 insects weight 1g and up to a million were caught in one evening.

The subsequent analysis gave the following results:

	%	
Loss at 110 °C	4.39	
Ash	5.64	
Protein	60.68	
Chloroform-soluble fraction	11.71	⎫
		⎬ polythenoid fatty acids
Alcohol – after chloroform extract	17.78	⎭
	100.20	

During their nuptial flights the eggs of these insects are laid into the water, the larvae live in the mud. The pupae and adults do not feed; larvae feed on organic matter (including algae) in the mud and 'form an important source of food material for (some) fishes, the composition of the fat of which is closely similar to that of the insects, which in turn resembles that present in algae but different from the composition of most terrestrial insects hitherto recorded'. The emergence of these, as of other mass appearing species, constitutes a considerable loss of protein and fat from the river economy.

Remarks on some benthos features have been made in other chapters: the molluscs of Nubia (ch. 4a), where the subfossil and recent fauna show similarities and differences; in ch. 16 the killing of large 'oyster beds' of *Etheria* by the new sediments of the Roseires basin are mentioned. Faunal lists and zoogeographical speculations on some aquatic groups are given in ch. 6d; the impact of water-bred insects on the demography of human populations is treated in ch. 23.

The formation of a new bottom fauna in the great basin of the Aswan High Dam will be a matter for future investigations of both practical and fundamental value. The crabs of the Nile system are added here as separate part written by an expert.

REFERENCES

(Only some of the papers by P. S. Corbet and those by A. Tjønneland are quoted).

Corbet, P. S. 1964. Temporal patterns of emergence in aquatic insects. Can. Ent. 96: 264–279.

Corbet, P. S. & Tjønneland, A. 1955. The flight activity of twelve species of East African Trichoptera. Univ. Bergen Årb. no. 9: 49 pp.

Fryer, G. 1960. The feeding mechanism of some Atyid prawns of the genus *Caridina*. Trans. Roy. Soc. Edinb. 64: 217–244.

Fryer, G. 1961. The developmental history of *Mutela bourguignati* (Ancey) Bourguignat, (Mollusca Bivalvia). Phil. Trans. Roy. Soc. Edinb. B. 244: 259–298.

Grindley, D. N. 1952. The composition of the body fat of small green Chironomids. J. Exp. Biol. 20: 440–444.

· Lewis, D. J., Henry, A. J. & Grindley, D. N. 1954. Daily changes in the numbers of Chironomid midges at Khartoum. Proc. Roy. Ent. Soc. Lond. 29: 124–128.

Lewis, D. J. 1957. Observations on Chironomids at Khartoum. Bull. Ent. Res. 48: 155–184.

MacDonald, W. W. 1953. Lake – Flies. Uganda J. 17, Kampala.

MacDonald, W. W. 1956. Observations on the biology of Chaoborids and Chironomids in Lake Victoria and on the feeding habits of the 'Elephant-Snout fish' *(Mormyrus kannume* Forsk.). J. Anim. Ecol. 25: 36–53.

Mandahl-Barth, G. 1954. Freshwater molluscs of Uganda and adjacent territories. Ann. Mus. Roy. Congo Belge, 8.

Monakov, A. V. 1969. The zooplankton and zoobenthos of the White Nile and adjoining waters in the Rep. of the Sudan. Hydrobiologia 33: 161–185.

Rzóska, J. 1964. Mass outbreaks of insects in the Sudanese Nile basin. Verh. Int. Ver. Limnol. 15: 194–200.

Rzóska, J. 1974. The Upper Nile swamps, a tropical wetland study. Freshw. Biol. 4: 1–30.

Thomas, I. F. 1961. The Cladocera of the swamps of Uganda. Crustaceana 2: 108–125.

Tjønneland, A. 1960. The flight activity of mayflies as expressed in some East African species. Årbok Univ. Bergen, Mat. Naturv. 1: 1–88.

Wülker, W. 1963. Prospects for biological control of pest Chironomidae in the Sudan. Rep. WHO (WHO/EBL) 11: 1–25.

Wülker, W. 1963. Investigations on the Chironomid fauna of the Nile, (Khartoum, Wad Medani, Sennar, Wadi Halfa). Hydrob. Res. Unit, Univ. Khartoum. Ann. Rep. 9/10: 20–21.

25a. FRESHWATER CRABS OF THE NILE SYSTEM

by

T. R. WILLIAMS

The River-crabs (Potamidae) of the Sudan

In 1931 Major S. S. Flower brought together the results of 25 years interest in the Crustacea in his 'Notes on Freshwater Crabs in Egypt, Sinai and the Sudan' (Proc. Zool. Soc. Lond.). In doing so he expressed the hope that his paper might prove to be of use as a starting point for other workers. Unfortunately this wish cannot yet be said to have matured, as even today there is still little to add to Flower's engaging account of the Nile's river-crabs.

It was therefore with considerable interest that I received a collection of one of these crabs, *Potamonautes niloticus* (H. Milne Edwards), made by Mr. Hammerton from the Blue and White Niles at Khartoum. This striking and unmistakable species has the carapace ornamented with a series of lateral spines which presumably afford a measure of protection against predation. The ease with which it sheds its limbs on capture may serve the same purpose. *P. niloticus* tolerates a wide range of conditions, appearing as much at home in lakes as in stony or muddy bottomed rivers and streams, and it is perhaps the most widely distributed of the African Potamids with a range which extends from the Nile Delta to the affluent rivers of Lake Victoria. A particular point of interest concerning this crab is that it has been the subject of a physiological investigation which indicated that its distribution could be interpreted in terms of the influence on its ionic regulating mechanism of such factors as water temperature and concentration of dissolved salts (Shaw, J. 1959, J. Exp. Biol. 36).

In Kenya and Uganda *P. niloticus* is associated with the early stages of *Simulium neavei*, a vector of human onchocerciasis, and its distribution in this part of the Nile drainage area is in consequence extremely well known. In general it can be said that *P. niloticus* occurs in the lower reaches of all rivers other than those few which are unusually poor in salts. In these, and in the cooler upper reaches of all rivers, it is replaced by different species. Swamps waters and papyrus fringed lake shores are also unsuitable, presumably on account of oxygen depletion (Shaw 1959), while for reasons which have yet to be explained this species is unknown from Lakes Edward and George and their inflowing streams.

In the Sudan *P. niloticus* has been recorded from numerous points on

353

the Nile above Khartoum, from the Blue Nile as far south as Singa (though no doubt it extends further) and from the Sobat. It may also occur in the Sudanese Baro River, as Mr. A. W. R. McCrae has sent me a specimen from near Gambela, S. W. Ethiopia. A puzzling feature of its distribution has been, as Flower noted, the absence of records from the entire length of the White Nile and the Bahr el Gebel. Earlier this year (1969), however, Mr. Hammerton recovered a carapace from the river-bed at Kodok and it is therefore possible that *P. niloticus* is present, though rare, elsewhere in the White Nile. The absence of crab records from the Sudd fauna suggests that Flower was correct in believing this species to be absent from at least extensive reaches of the Bahr el Gebel. If this is indeed the case *P. niloticus* must have a disjunct distribution within the Nile system and it is therefore not surprising to find that minor, though distinct, morphological differences distinguish specimens from Khartoum from those collected in East Africa. Some of these differences have been noted by Pretzmann (Ann. Naturhistor. Mus. Wien 65, 1962), but to these may be added the more slender limbs and the smaller size at maturity of the Khartoum crabs. It is also possible that the Sudanese and East African populations differ physiologically, for at times potassium concentration in the Blue Nile falls below the limiting value discovered by Shaw (1959) in Uganda.

Still less can be added to Flower's accounts of the two other species he recorded from the Sudan. One of these, *P. floweri* (de Man), was first collected by Flower himself at Shambe and at several other points on the Bahr el Gebel. Its known distribution is still essentially as given by Flower in 1931, namely south from the Bahr el Gebel into Uganda and west across the Nile-Congo divide to the Ubangi River. (But a closely related, if not the same, species occurs in Gabon. Perhaps *P. floweri* extends right around the Congo basin?). An account of the habits of this species in the Uele district of the Congo shows it to be more freely terrestrial than *P. niloticus*, burrowing actively and apparently aestivating within its burrows (Rathbun, M. J. 1921. Bull. Amer. Mus. Nat. Hist. 43).

A particular difficulty with Potamidae is the general similarity of appearance of many quite distinct species, but the resulting taxonomic difficulties can nearly always be solved by reference to the form of the male genitalia. Unfortunately many species had been described before the value of the genitalia had been appreciated, among which was *P. berardi* (Audouin), the last of the Sudanese crabs to be mentioned by Flower. It seems unlikely that this species occurs far outside the lower (Egyptian) Nile valley, but at various times it has been recorded from such distant and widely separated localities as Ethiopia, Ruanda and the highlands of eastern Uganda. The true distribution of *P. berardi* is in fact very uncertain, but as it is probable that a record given by Flower from Wadi Halfa is correct it may be included in the Sudanese fauna. Nothing is known of

354

the habits of this crab apart from the fact that it always appears to have been taken in water and in many cases from canals.

In conclusion, a recent collection of crabs from the Imatong Mts. on the Sudan-Uganda border (A. W. R. McCrae) have proved to be a new and undescribed species. It may well be that other new species of crabs remain to be discovered from within the Sudan.

Freshwater crabs of the Nile system outside the Sudan and Egypt

Knowledge of this area ranges from the very detailed, where crab-associating Simuliidae transmit onchocerciasis, to completely absent. Highlands of Kenya and Uganda have been especially well worked, but much less is known of the Ethiopian fauna. At lower altitudes collecting has been extensive east and west of Lake Victoria, but scanty or totally lacking to the north and south. It is nevertheless clear that the three main highland areas (western Uganda, eastern Uganda and western Kenya, and Ethiopia) have distinctive faunas, and that differences also exist between 'lowland' areas east and west of L. Victoria.

Major lakes and adjacent areas

Crabs are not recorded from L. George and surprisingly, and despite searching (A. W. R. McCrae, in litt.), none are known from L. Edward. The almost ubiquitous Nile crab, *Potamonautes niloticus* (H. Milne Edwards), occurs in L. Albert and L. Victoria, and extends for considerable distances into inflowing rivers. A few rivers entering L. Victoria from the north and east have *P. gerdalensis* Bott. This is probably Shaw's (1959) undetermined crab which replaces *P. niloticus* in ion-deficient rivers.

A very variable species or species group is known from L. Victoria and from moderate altitudes mainly in western Uganda. Included taxa are *P. emini* (Hilgendorf) from L. Victoria, *P. mutandensis* (Chace) from Lakes Mutanda and Bunyoni, and, perhaps, *P. amalerensis* (Rathbun) from Mt. Debasien. These 'emini group' crabs also appear to be widespread in swampy rivers between Lakes Victoria and Albert. While their distribution cannot be fully known there seems to be no doubt that they are absent to the northeast of L. Victoria. The reverse is true of *P. gerdalensis* which has not been taken in the western rivers.

Highlands east of Lake Victoria

Between 5,000 and 6,000 ft. (1500–1800 m) on Mt. Elgon and in western Kenya *P. niloticus* is replaced by *P. loveni* (Colosi). Also restricted to this area is an unnamed species, smaller than *P. loveni*, but closely related to it. (Hynes, Williams & Kershaw 1961, Williams, Hynes & Kershaw 1964).

355

In upper reaches around Lake Albert *P. niloticus* is succeeded by *P. aloysiisabaudiae* (Nobili), which is also the common crab of the Kigezi and Ruwenzori rivers from which *niloticus* is absent (Williams, et al. 1964). *P. idjwiensis* (Chace), previously known only from L. Kivu, is now also known to occur on the Ruwenzori, where it must be extremely rare. Its interest is that its affinities appear to be with *P. loveni* and provide the only evidence of a link between the highland crab faunas of east and west Uganda.

ETHIOPIA

Two large species, *P. antheus* (Colosi and *P. ignestii* (Parisi), are definitely known from the Nile region, the former widely distributed, the latter apparently not. Their affinities are interesting; *P. antheus* is closely related to *P. aloysiisabaudiae*, *P. ignestii* is less certainly akin to *P. berardi* (Egypt).

The smaller Ethiopian crabs are not at all well known, but have something in common with those of the Imatong Mts. This reinforces the connection with western Uganda shown by *P. antheus*, and further emphasises the isolation of the Mt. Elgon-Kenya Highland fauna.

REFERENCES

Flower, S. S. 1931. Notes on freshwater crabs in Egypt, Sinai and the Sudan. Proc. Zool. Soc. Lond. 1931: 729–735.

Hynes, H. B. N., Williams, T. R. & Kershaw, W. C. 1961. Freshwater crabs and *Simulium neavei* in East Africa. Ann. Trop. Med. Parasit. 55: 197–201.

Rathbun, M. J. 1921. The Brachyuran crabs collected by the American Museum Congo expedition. Bull. Amer. Mus. Nat. Hist. 43: 379–468.

Shaw, J. 1959. Salt and water balance in the East African crab *Potamon niloticus* (M. Edw.). J. Exper. Biol. 36: 157–176.

Williams, T. R. 1959. The diet of freshwater crabs associated with *Simulium neavei* in East Africa. I. Crabs from Western and Eastern Uganda collected by the Cambridge E. Afric. Expedition 1959. Ann. Trop. Med. Paras 55: 128–131.

Williams, T. R., Hynes, H. B. N. & Kershaw, W. E. 1961. Maintenance and diet of African freshwater crabs associated with *Simulium neavei*. Ann. Soc. Belge Med. Trop. 41: 291–292.

26. WATER CHARACTERISTICS

by

J. F. TALLING

Introduction

Two characteristics of water from the lower Nile long attracted attention – that it was of excellent quality for drinking and irrigation, and that it seasonally contained high concentrations of silt. The systematic, scientific exploration of the general subject developed but slowly over the last hundred years. An essential element in this was a direct parallel to geographical exploration, namely the tracing upstream of sources of the materials and properties observed below. Geography and geology dictate that a blending of very different types of water must occur in the head-waters and middle reaches, with contributions from dissimilar lakes and highland tributaries. These sources also have different seasonal flows, and so influence the seasonal composition of water downstream. Further seasonal influences include the annual impoundment of water by dams in the Sudan, and the chemical activities of plankton in the impounded water. Unlike many rivers, the Nile is little affected by pollution. However, the extensive Sudd swamps on the Upper White Nile can introduce deoxygenation and its chemical consequences.

As in other fields of river ecology, the information is obtained by combining repeated (e.g. seasonal) observations at a few fixed sites with longitudinal surveys. In lower Egypt, analyses of Nile water extend back at least 100 years; much early information is given by Lucas (1908). Later descriptions of seasonal changes near Cairo include Aladjem (1926, 1928), Abdin (1948a), Hurst (1957), and Ramadan (1972). At Khartoum, careful work by Beam (1906, 1908) provided detailed seasonal analyses of the lower White and Blue Niles, and also a foundation for interpreting the important longitudinal changes down the White Nile. The latter were explored more recently by Talling (1957a), Bishai (1962), and Kurdin (1968), as were seasonal changes in the two rivers at Khartoum (Prowse & Talling 1958; Talling & Rzóska 1967; Abu Gideiri 1969; Omer Badri 1972; Hammerton 1972d). Chemical and physical studies of the headwater lakes were developed in part by hydrological expeditions from the lower Nile (e.g. Hurst 1925; Grabham & Black 1925; Tottenham, 1926), in part independently of other work on the Nile; they are summarised by Talling & Talling (1965) and Beadle (1974). Much information on water characteristics can also be found in various regional chapters of this book. The water chemistry of the Nile is discussed by Golterman (1975) in relation to general aspects of river chemistry.

Temperature

This factor has a double significance. Absolute values (and their seasonal ranges) are of importance for chemical reaction rates, equilibria, and biological tolerances; vertical differences can regulate, through water density, the overall stratification of a water-column.

Over most of the Nile a prolonged or seasonal thermal stratification is absent, due to wind- and current-induced mixing in shallow waters. Among the headwater lakes, this is applicable to the relatively shallow L. Tana (Morandini 1940) and L. George (Viner & Smith 1973), and was once believed true for the deeper L. Albert also (Worthington 1930). However, periods of stratification are probably important in the economy of L. Albert (Talling 1963; see Fig. 96), though the vertical temperature differences are small. In Lake Victoria, an annual cycle of thermal stratification can be distinguished (Fish 1957; Talling 1957c, 1966; see chapter 10), and a pronounced thermocline is present – at least seasonally – in L. Edward (Beadle 1932, 1974; Damas 1937). Thermal stratification appears briefly and irregularly in the downstream reservoir at Roseires (Hammerton 1972b, c; see chapter 16), but is annual and prolonged in Lake Nubia – Nasser (see chapter 19a). Prolonged stratification is lacking in the shallower reservoirs of Gebel Aulia and Sennar in the Sudan, and – according to Abdin (1948b) – was absent from the old Aswan reservoir.

Even though the shallower lakes and reservoirs have frequent vertical mixing, short *diurnal* phases of thermal stratification are widespread and important. They have been studied in detail for the Gebel Aulia reservoir (Talling 1957b), some inshore areas of Lake Victoria (Worthington 1930; Talling 1957b), and Lake George (e.g. Viner & Smith 1974; Ganf 1974b). Diurnal changes of stratification in Lake Tana have also been described by Bini (1940) and Morandini (1940). Fig. 85 shows an example from Gebel Aulia, in which the daily stratification of temperature, oxygen and pH was broken down at night by a combination of surface heat loss and wind action. Further implications were traced by Talling (1957b) for the depth-distribution of a planktonic blue-green alga, *Anabaena flos-aquae f. spiroides*, and the estimation of photosynthetic productivity (see chapter 27).

The temperatures reached depend upon factors influencing the heat (energy) budget, including radiation income (solar elevation, day length, cloudiness), surface exchange of sensible heat (air temperature), back-radiation and latent heat of evaporation. The last two components are large and increase markedly under conditions of low relative humidity, as found in the central and northern Sudan and in Upper Egypt. There, as

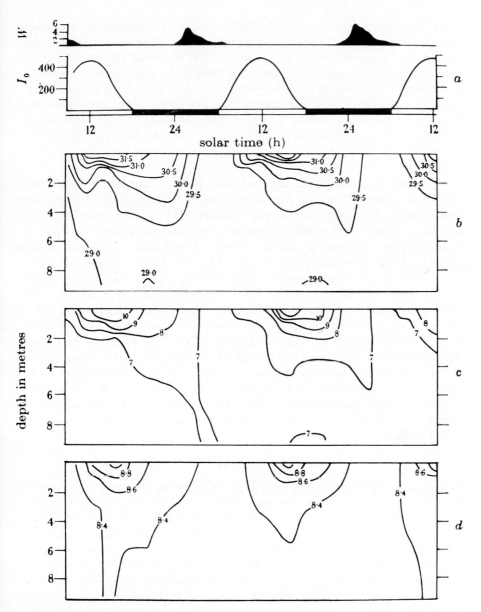

Fig. 85. Diurnal changes in the Gebel Aulia reservoir, 6–8 October 1955, of (a) wind strength W (as approx. force numbers, Beaufort scale) and incident solar irradiance I_0 (in kerg/cm² · s = W/m²J, spectral region 400–700 nm), (b) temperature (isotherms in °C), (c) dissolved oxygen (isopleths in mg/l), and (d) pH. (from Talling 1957b).

Entz (Chapter 19a) notes, the surface water temperatures are often considerably below the mean air temperatures. The evaporation component has been evaluated by Gorgy (1959) for Lake Qarun (Faiyum) from an estimated energy balance. Although the evaporation loss from Lake Nasser has engaged much attention (e.g. Omar & El Bakry 1970), published calculations of the energy balance seem to be lacking for the Nile itself. For the different circumstances of Lake Tana, Hutchinson (1957, p. 522) has used data of Morandini (1940) to estimate the various contributions to nocturnal heat loss, with evaporative cooling predominant.

Since the Nile flows through a wide belt of latitude, longitudinal (N–S) gradients of water temperature might be anticipated. These are best shown (Fig. 86) during the 'winter' (Dec.-Jan.) season of the northern areas, when water temperatures from Khartoum (Prowse & Talling 1958; Talling & Rzóska 1967; Abu Gideiri 1969) to Cairo (Abdin 1948a; Ramadan 1972 – see Fig. 89) fall to about 17–19 °C, in contrast to more maintained levels near 28 °C in the southern Sudan (Talling 1957a; Bishai 1962). The headwaters are typically a little cooler, due to altitude and probably the reduction of insolation by cloud cover. Talling (1969, Fig. 2) has compared the relatively constant surface temperatures of lakes Victoria (23.9–27.1 °C), Edward, and Albert with the large seasonal range at Aswan (16–30 °C: from Abdin 1948b), as part of general seasonal trends in African freshwaters. The maximum temperatures in Lake Nubia-Nasser are among the highest in large African lakes. Chapter 19a gives an outline of the variability of temperature in this river-lake, which is seasonally affected by a massive, cooler, inflow.

Transparency and light penetration

As in many freshwaters, and probably most rivers, these related features are mainly controlled by the particulate content of the river water. The more numerous measurements of transparency denote the depth at which a white, or black and white, disc (Secchi disc: diameter usually 20 or 30 cm) disappeared from view. This inverse measure of turbidity is more strongly influenced by the scattering of light than is light penetration (or extinction) itself. The latter is best estimated, for specific spectral regions isolated by colour filters, from the response of a submerged photo-cell which – in homogeneous water – decreases exponentially with depth. The rate of decrease is measured by the vertical extinction (attenuation) coefficient (ε), which is expressed in ln units/m and varies with wavelength. The depth with 1% of the surface response ($z_{1\%}$ in m) is related to ε by the relationship

$$z_{1\%} = \frac{4.6}{\varepsilon}$$

The minimum extinction coefficient over the visible spectrum, ε_{min}, usually dictates the depth of the euphotic zone (z_{eu}) and photosynthetic zone. An approximate relationship (Talling, 1965) is

$$z_{eu} \simeq \frac{3.7}{\varepsilon_{min}}$$

Many measurements of light penetration in the Nile have been made without colour filters, over a broad spectral region (e.g. Abdin 1948a, b, c), and the corresponding extinction coefficients are slightly higher than ε_{min}. The latter can be used as a summarizing characteristic, as can $z_{1\%}$.

The clearest water investigated in the Nile system is that of offshore Lake Victoria (Levring & Fish 1956; Fish 1957; Talling 1957c, 1965, 1966). The measured variation over a year of ε_{min} at one station was 0.16–0.31 ln units/m ($z_{eu} = 12$–21 m), with higher extinction at times of higher phytoplankton density (Talling 1966). Light penetration is much lower in the other headwater lakes (Albert, Edward, George, Tana); the most penetrating spectral region shifts from green, through orange, to red in the turbid waters of L. Tana and L. George. High extinction in the latter is chiefly governed by the dense phytoplankton which discolours the water (Ganf 1974a). In lakes Edward and especially Albert (Levring & Fish 1956; Talling 1965) both phytoplankton and fine silt contribute to the extinction, and silt is probably dominant in L. Tana – for which a single measurement near Bahar Dar exists (Talling unpublished, March 1964: $\varepsilon_{min} = 1.2$ ln units/m, $z_{eu} \simeq 3$ m). The Secchi transparency is usually about one-third to one-half of the euphotic depth.

In the flowing river water, more detritus is carried in suspension and light penetration is usually low. Thus a longitudinal section of the White Nile in June 1954 (Fig. 86) yielded ε_{min} coefficients of minimum extinction – displaced to the red spectral region – between 5 and 11 ln units/m, with transparency between 0.1 and 0.4 m. Higher values of transparency in the following December indicate considerable temporal variation, probably related to water level and the seasonal flow of some mountain torrents (Talling 1957a). Seasonal changes are accentuated in the Blue Nile, with transparency reduced to virtually zero by the load of silt carried in the flood water (Fig. 94). This silt has produced a seasonal minimum of transparency (and $z_{1\%}$) downstream, as at Aswan (Abdin 1948b, c) and Cairo (Abdin 1948a, Ramadan 1972).

The storage of water in the Nile reservoirs involves two opposing influences on the particulate content, and so on light penetration and transparency. Larger particles will sediment out, but the enhanced growth of phytoplankton may reduce light penetration. Net clarification seems characteristic of the old Aswan reservoir (Abdin 1948b; Elster & Vollenweider 1961) and the present Lake Nasser (Entz 1972, and chapter

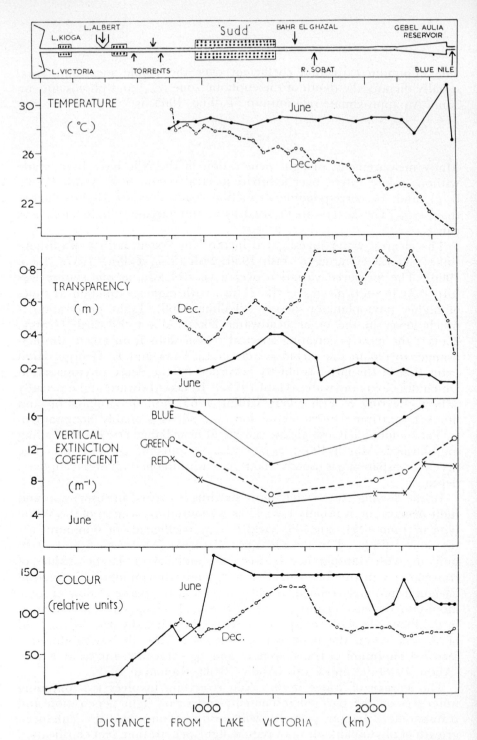

Fig. 86. The variation of some physical properties – temperature, transparency, light extinction, and colour – along the White Nile between Lake Victoria and Khartoum, in June and December 1954. See also map in Fig. 87. (from Talling 1957a).

362

19), with Secchi transparency values often reaching several metres and exceptionally 6 m in upwelling situations. The new Roseires reservoir has also led to increased transparency in the lower Blue Nile (Chapter 16) as well as in the reservoir itself (Fig. 58). However, phytoplankton can greatly reduce transparency in L. Nasser (Entz 1972, and Chapter 19), and its effect appears dominant in most longitudinal sections of the filled Gebel Aulia reservoir (Figs. 86, 92, and Bishai 1962; Kurdin 1968). A longitudinal section from Aswan to Cairo (Elster & Vollen-weider 1961) showed a decrease of transparency downstream, probably due to increase of silt and detritus in flowing water. In the Sudd swamp region, where many measurements have been made (see Rzóska 1974), optical conditions in flowing and standing waters are not sharply con-trasted; the former typically carry much detritus, and the latter abundant phytoplankton. Unusually high transparency exists in the Bahr el Ghazal tributary, reaching 2.5 m or more.

Some biological implications of the varying light penetration are mentioned in Chapter 27, including the restricting influence of high 'background' absorption upon primary production. The increased transparency brought about downstream by the new Roseires dam has had an adverse effect upon a traditional style of fishing by cast-nets (Hammerton 1972d). In the Gebel Aulia reservoir, no well-defined pattern of diurnal vertical migration by zooplankton could be demon-strated (Rzóska 1968), possibly due in part to the shallow euphotic zone affected by wind-induced turbulence.

Salinity and conductivity

As in most freshwaters, salinity and discharge tend to be inversely related. Their relationship also depends upon the balance between precipitation and evaporation, and the prevalence in the drainage basin of readily-eroded and -weathered minerals and of exposed saline deposits (e.g. evaporites) from earlier geological periods. The 'Amazonian' combination of very high discharge and very low salinity is unknown in the Nile system, where even the flood water of the Blue Nile has an ionic content greater than 1 meq/l of cations or anions and electrical conductivity (20 °C) above 100 μmho/cm. However, the ionic content and conductivity fall appreciably in the seasons and reaches affected by flood water from the Blue Nile and River Atbara, as at Khartoum (Figs. 88, 94), Lake Nasser (Entz; Chapter 19a), and Cairo (Aladjem 1926, 1928; Hurst 1957; Ramadan 1972: see Fig. 89). Entz has shown that in Lake Nasser the conductivity differences can be used to identify water masses and trace their movement downstream. Analyses made before the High Dam showed that there was an appreciable increase in solutes between Aswan and Cairo (Aladjem 1928; Elster & Vollen-weider 1961).

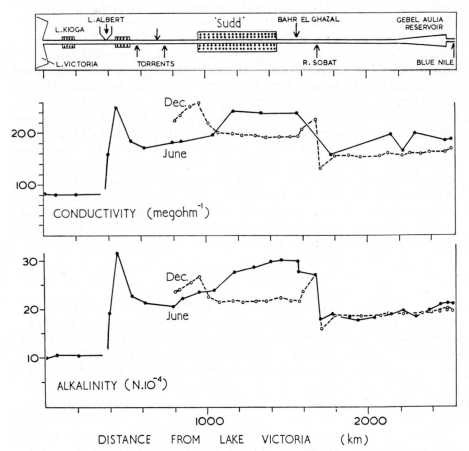

Fig. 87. The variation of conductivity (megohm^{-1} = μmho/cm, 20°C) and alkalinity (N. 10^{-4} = meq/l × 10) along the White Nile between Lake Victoria and Khartoum, in June and December 1954. (from Talling 1957a).

It is in the headwaters that the greatest variation of salt content occurs. The existence there of some relatively saline waters was established by early surveys (Hurst 1925, 1957; Tottenham 1926) with gravimetric analyses, but most results are based upon estimations of conductivity (see Talling & Talling 1965). This increases ninefold along the lake series Victoria – Tana – Albert – Edward (Tables 1, 2), and it is the lesser contribution of high conductivity water from L. Albert and the Western Rift which raises the conductivity of the White Nile from the low initial level of L. Victoria (Fig. 87). As the volume derived from L. Albert can vary over short periods, (Tottenham 1926), owing to wind action and the peculiar Nile inflow–outflow configuration (Fig. 31), conductivity can fluctuate in the Albert Nile (Beauchamp 1956) and pulses of higher conductivity water are probably still recognisable in the Sudan (Talling 1957a). Further downstream, an increase can sometimes occur in the Sudd swamp region where much water is lost by evapo-transpiration. A diluting effect of the River Sobat is more consistently found (Fig. 87; see also Rzóska 1957, Bishai 1962), and which will be greater during the seasonal high flows in this river. At a time of low levels here and at Khartoum, in March-April 1958, Gay (1958) found high conductivity values near 350 μmho/cm (20 °C) along a considerable length of the White Nile. As values in the same season of 1954–1956 and one later year were much lower (Talling unpublished; Abu Gideiri 1969), sporadic increases are suggested, possibly related to the saline discharge from Lake Albert. An exceptionally low conductivity of 40 μmho/cm was encountered by Rzóska (1957, 1974) at Lake Ambadi on the Bahr el Ghazal, whose waters are distinctive in other respects.

On the Blue Nile, there is a net increase of conductivity in the Gorge region between Lake Tana and Roseires (Table 2). Little further change was found by Talling & Rzóska (1967) and Hammerton (1970) to Khartoum, where values are still lower than in the White Nile – and especially so in the flood season.

In the Delta lakes (chapter 20), the salt content is increased by penetration of sea-water as well as by evaporation and local pollution (Elster & Vollenweider 1961; Aleem & Samaan 1969; El-Wakeel & Wahby 1970). Thus chlorinity is a convenient index of the salinity. Older values were usually in the range 0.5–5‰ Cl, but increases have occurred since the High Dam altered the hydrological and silting regimes. At the same time, the previously marked effects of the annual Nile flood water upon salinity (and fisheries) of the south-eastern Mediterranean (cf. Hecht 1964; Oren 1969) have disappeared.

Major ions

The bicarbonate plus carbonate ions strongly dominate the total anionic content in virtually all Nile waters (Delta lakes excluded). As their combined concentration is closely approximated by the titration alkalinity – the effects of H^+, OH^-, and $SiO(OH)_3^-$ being generally negligible – alkalinity is very closely correlated with total ionic concentration and so with conductivity. Approximate relationships, applicable to many

365

other African freshwaters (Talling & Talling 1965), are

conductivity (μmho/cm, 20°C) $\simeq$ 100 x (alkalinity, meq/l)

$\simeq$ 85 $\times$ (total cations or anions, meq/l)

For examples, see Tables 1, 2 and Figs. 87, 90, and 94. Numerous and relatively precise measurements of alkalinity are available for waters throughout the Nile system, and give useful indications of total ionic content. The general range for alkalinity is roughly 1–3 meq/l in the major stretches of the river, with an upper limit of 9 meq/l in L. Edward.

The remaining major anions, Cl⁻ and SO_4^{2-}, show different patterns of distribution. Concentrations of chloride are very low (∼0.05–0.15

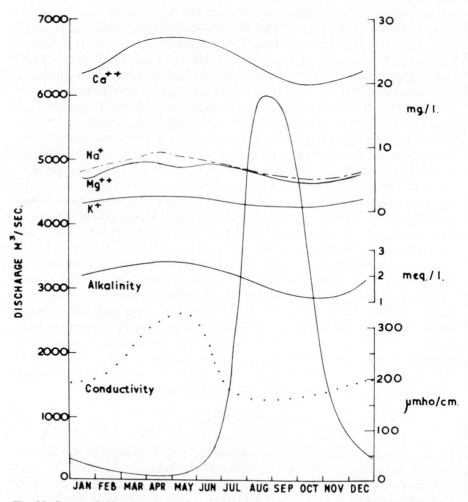

Fig. 88. Seasonal changes in concentrations of major cations, alkalinity, and conductivity (20°C) in the Blue Nile at Khartoum, shown in relation to discharge. (from Hammerton 1972d).

366

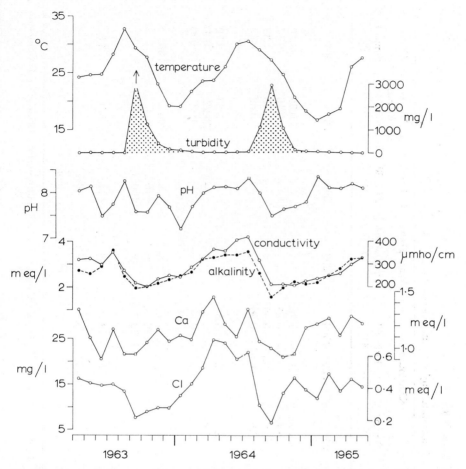

Fig. 89. Seasonal changes of seven physical and chemical characteristics of Nile water at Cairo, before the High Dam. Turbidity maxima indicate the annual flood water. (adapted from Ramadan 1972).

meq/l) in the major headwaters unaffected by the saline injection from the Western Rift, but are much increased below this in the White Nile (Talling 1957a). Further downstream, the River Sobat has a strong diluting influence, as has the Blue Nile (Fig. 88) for the combined waters in the Main Nile below Khartoum. The extremely low concentrations described for the White Nile by Bishai (1962) are incompatible with all other analyses. Concentrations measured at Aswan and Cairo (Table 1, Fig. 89) indicate additional inputs to the river in Egypt, although a seasonal minimum existed during the flood-water of the pre-High Dam era (Fig. 89; see also Hurst (1957) for 1933–1936 analyses).

Sulphate is present in low concentration (~ 0.05 meq/l) in the two headwater lakes of Victoria and Tana. Some earlier and still lower

367

Table 1. Water characteristics of the White and Main Niles

Characteristic	Lake Victoria 2.iii.61 (1)	nr. Lado (S. Sudan) 25.xii.54 (2)	Place, date, and reference (footnote) Khartoum 19.iii.63 (3)	Khartoum 1965–67 (4)	Wadi Halfa 15–20.iii.64 (5)	L. Nubia 23.ii.68 (5)	L. Nasser nr. Aswan 23.iii.74 (6)	Cairo 29.iii.74 (6)
conductivity (k_{20}) μmho/cm	97	235	240	220 –500	290	350	237	307
as meq/l:								
Na^+	0.45	—	1.26	1.09– 1.78	—	—	0.85	1.20
K^+	0.10	—	0.18	0.18– 0.24	—	—	0.16	0.17
Ca^{2+}	0.28	0.38	0.92	0.71– 0.95	—	—	1.18	1.39
Mg^{2+}	0.21	—	0.65	0.62– 0.96	—	—	0.62	0.82
$HCO_3^- + CO_3^{2-}$ (alky)	0.92	2.40	2.80	2.3 – 3.3	—	—	2.33	2.70
SO_4^{2-}	0.05	0.10	0.04	—	—	—	0.22	0.37
Cl^-	0.11	0.28	0.14	0.17– 0.27	—	—	0.16	0.41
as μg/l:								
total P	47	—	—	—	—	—	90	49
$PO_4 \cdot P$	13	60	—	2 –120	20	10	—	—
$NO_3 \cdot N$	11	30	—	10 – 90	'0'	100	—	—
as mg/l:								
Si($\times 2.1 = SiO_2$)	2.0	2.8	13.5	8.4 – 13	10	11	6.9	4.8
pH	8.0	7.9	—	8.0 – 8.9	8.1	8.4	—	—
transparency (m)	—	0.50	—	—	0.20	0.98	—	—

(1) Talling (1966), Talling & Talling (1965); (2) Talling (1957a); (3) Talling & Rzóska (1967); (4) Hammerton (1972d); (5) Hammerton (1972b); (6) Talling & Heron, unpubl.

Table 2. Water characteristics of the Blue Nile.

Place, date, and reference (footnote)

Characteristic	Lake Tana 8.xii.65– 18.iv.66 (1)	11– 12.iii.64 (2)	Tissisat Falls 12.iii.64 (2)	Gorge, M. bridge 10.iii.64 (2)	10.iv.64 (2)	Khartoum 1965–67 (3)	Roseires reservoir xi.67 (4)	v.68 (4)
conductivity (k_{20}, μmho/cm)	200 –240	137	156	231	238	140 –390	200	300
as meq/l:								
Na^+	0.31– 0.41	0.24	0.26	0.57	0.47	0.20– 0.39	0.19	0.36
K^+	0.03– 0.05	0.04	0.05	0.07	0.06	0.04– 0.07	0.04	0.06
Ca^{2+}	0.70– 0.76	0.94	0.93	1.52	1.50	0.98– 1.41	1.03	1.27
Mg^{2+}	0.50– 0.70	0.45	0.52	0.80	0.72	0.41– 0.54	0.42	0.56
$HCO_3^- + CO_3^{2-}$ (alky)	1.52– 1.92	1.52	1.57	2.23	2.57	1.63– 2.66	1.74	2.46
SO_4^{2-}	—	0.05	0.08	0.48	0.36	—	—	—
Cl^-	0.03– 0.07	0.04	0.05	0.11	0.08	0.06– 0.21	0.10	0.20
as μg/l:								
$PO_4 \cdot P$	15 – 20	30	24	24	—	2 –120	35	<4
$NO_3 \cdot N$	0 – 80	—	—	—	—	1 –100	55	2
as mg/l:								
$Si(\times 2.1 = SiO_2)$	8.4 – 9.4	6.8	6.8	8.9	8.7	7.5 – 11	9.4–11	
pH	8.1 – 8.4	8.4	8.5	—	—	8.2 – 9.1	—	—
transparency (m)	—	1.1	—	—	—	—	—	—

(1) Hammerton (1972b); (2) Talling unpublished, Talling & Rzóska (1967); (3) Hammerton (1972d); (4) Hammerton (1972b).

369

Table 3. Water characteristics of Nile tributaries.

Characteristic	Lake Edward 22.vi.61 (1)	Lake Albert xi.60–viii.61 (1)	Bahr el Ghazal 30.xii.53 (2)	Bahr el Ghazal 4.xii.60 (3) ←	River Sobat xii.07–x.08 (4)	River Sobat 10.xii.54 (2)	River Sobat 1.i.61 (3)	River Sobat 2.v.64 (5) →	R. Atbara vi–x.07 (6)
conductivity (k_{20}, μmho/cm)	925	720 –780	—	100	—	112	110	—	—
as meq/l:									
Na^+	4.78	3.96	—	—	0.01– 0.31	—	—	—	—
K^+	2.30	1.66	—	—	<0.01– 0.13	—	—	—	—
Ca^{2+}	0.62	0.45– 0.54	—	—	0.36– 0.59	0.44	—	—	—
Mg^{2+}	3.93	2.57– 2.64	—	—	0.23– 0.56	—	—	—	—
$HCO_3^- + CO_3^{2-}$ (alky)	9.85	7.27– 7.33	2.14	1.06	0.47– 1.01	1.52	1.00	1.15	2.0–2.8
SO_4^{2-}	0.65	0.56– 0.94	—	—	'none'	<0.03	—	—	<0.02
Cl^-	1.01	0.93– 1.08	<0.06	—	<0.03– 0.05	<0.06	—	—	0.05–0.12
as μg/l:									
total P	127	120 –200	—	—	—	—	—	—	—
$PO_4 \cdot P$	18	120 –170	20	25	—	45	63	4	—
$NO_3 \cdot N$	24	4 – 33	—	25	—	15	80	160	—
as mg/l:									
$Si (\times 2.1 = SiO_2)$	3.0	0.02– 0.52	18	12	3.6 –10.7	12	14	—	—
pH	9.1	8.9 – 9.0	7.8	7.4	—	6.8	7.8	7.15	—
transparency (m)	—	—	1.2	1.6	—	0.35	0.45	0.30	—

(1) Talling (1963); (2) Talling (1957a); (3) Bishai (1962); (4) Beam (1911); (5) Kurdin (1968); (6) Beam (1908).

estimations (Beauchamp 1953) were probably influenced by an analytical error (cf. Talling & Talling 1965). Considerable enrichments occur from the Western Rift (L. Albert) discharge to the White Nile, and in the Gorge region to the Blue Nile (Tables 1, 2). Concentrations in the Blue Nile persist downstream in the Sudan plain, but in the White Nile a remarkable loss occurs as the water traverses the Sudd swamps. First discovered by Beam (1906, 1908), the effect is presumably due to microbial reduction.

As regards the major cations, the chief interest lies in the widely varying ratio of the divalent (Ca^{2+}, Mg^{2+}) to the monovalent (Na^+, K^+) ions. The Blue and White Niles contrast in this respect: calcium is predominant in the former, and sodium in the latter. The ratio downstream is particularly influenced by flood water from the Blue Nile, as noted at Cairo by Abdin (1948a). The fewer analyses of potassium show a particularly high content in the White Nile, which is probably derived largely from the potassium-rich waters of Lake Albert and the Western Rift drainage. Although the precipitation of $CaCO_3$ may well have influenced ionic proportions in the more saline headwater lakes of Edward and Albert (Talling & Talling 1965), there is no evidence for present-day losses of calcium on a large scale in the river itself, or in Lake Nasser (Entz, Chapter 19a). However, Hammerton (1970) found examples of benthic calcium carbonate deposition on a small scale in the Blue Nile.

Dissolved gases and pH

Despite possibilities for equilibration with the atmospheric reservoir, the content of dissolved oxygen and carbon dioxide may be altered by intense biological activity in several regions of the Nile system. Concentrations of the two gases are inversely correlated, if free rather than total carbon dioxide is considered, as is usual in natural waters. The total carbon dioxide content is dominated by the bicarbonate ion, which also accounts for the major part of the titration alkalinity, in turn cross-correlated with conductivity and salinity (see above).

Oxygen depletion is pronounced in two types of situation – swamp waters, and the deeper layers of stratified lakes and reservoirs. The first is illustrated by the passage of the White Nile (Bahr el Jebel) through the Sudd swamps (Talling 1957a; Bishai 1962; Kurdin 1968; Rzóska 1974), where a partial deoxygenation occurs (Fig. 90). Its extent appears to vary seasonally, as does the rapidity – and hence distance – of recovery downstream. Some important chemical consequences – affecting SO_4^{2-}, Fe, and possibly $PO_4 \cdot P$ – are noted in this chapter. Although less studied, severe depletions probably occur from time to time in swamp-affected regions of the River Sobat and its tributaries, and perhaps lead to fish kills there (Bacon 1918) as near Shambe on the Bahr el Jebel.

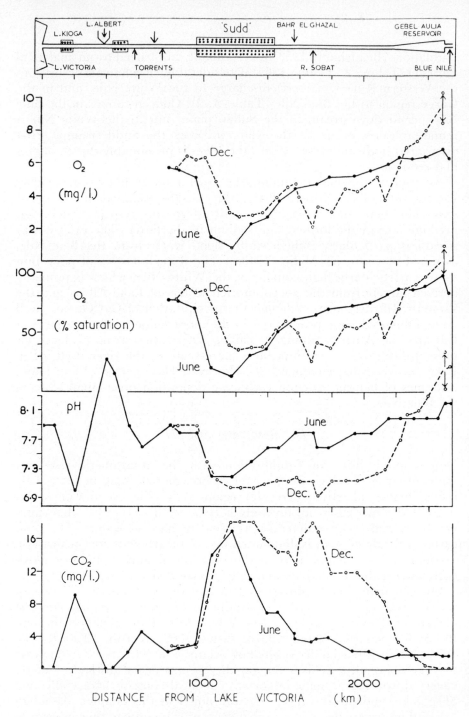

Fig. 90. The variation of dissolved oxygen, pH, and free carbon dioxide along the White Nile between Lake Victoria and Khartoum, in June and December 1954. See also map in Fig. 87. (from Talling 1957a).

The second, deep water, situation occurs annually in the stratified waters of lakes Victoria (Fish 1957; Talling 1966, see Chapter 10) and Nasser (Entz, Chapter 19a), for most of the year in Lake Edward (cf. Beadle 1974), and diurnally in Lake George (Ganf 1974b). The examples of deoxygenation in Lake Albert probably occur irregularly in time; they appear to develop from profile-bound density currents (Talling 1963; see Fig. 96), and may be responsible for the occasional kills of Nile perch *(Lates niloticus)* in this lake. In other Nile reservoirs, long periods of stratification with deep deoxygenation are lacking. However, short periods of variable deoxygenation have occurred in the Roseires reservoir, including an episode of complete deoxygenation soon after its first filling (Hammerton 1972a, b, and chapter 16). Extensive barriers to atmospheric exchange were introduced to parts of the White Nile by floating mats of *Eichhornia* (chapter 21).

Oxygen supersaturation, due to photosynthetic activity, is often encountered in regions with abundant phytoplankton. It develops diurnally and interacts with the diurnal temperature (density) stratification, as shown by Talling (1957; see Fig. 85) for the Gebel Aulia reservoir and by Hammerton (Fig. 91) for the White Nile at Khartoum. The latter involved a very wide, slowly flowing water-mass which differentiated both vertically and horizontally. In the neighbouring Blue Nile, surface supersaturation during the annual phytoplankton maxima was also found (Talling & Rzóska 1967). Further downstream, some very high levels of supersaturation have been recorded for some occasions and localities in Lake Nubia-Nasser (Entz, Chapter 19a). At Cairo, in the

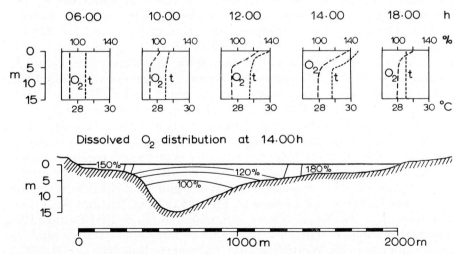

Fig. 91. Stratification in the White Nile near Khartoum. Above, diurnal changes of temperature and dissolved oxygen (as % saturation); below, cross-section of the river showing horizontal and vertical differences in dissolved oxygen. (Hammerton, unpublished).

373

pre-High Dam period, slight supersaturation developed during the cooler winter season and values were above 70% saturation throughout the year (Ramadan 1972).

Direct determinations of carbon dioxide content in the Nile system are few. They include swamp waters in Uganda (Carter 1955; Milburn & Beadle 1960) and Lake George (Ganf & Milburn 1971), which are characterised respectively by strong biological enrichment and depletion. Most estimates, of free CO_2, are deduced from pH and alkalinity. Though these are approximate, the large variations involved should be reasonably well-established. The inverse relation with oxygen concentrations is well shown by longitudinal surveys along the White Nile (Talling 1957a; Bishai 1962; Kurdin 1968), where CO_2 – enrichment occurs in the swamp regions and is reduced rather slowly downstream (Fig. 90). It is also shown by a comparison of the seasonal changes in areal contents of oxygen and free carbon dioxide in Lake Victoria (Talling 1966). Depletions of carbon dioxide, involving free CO_2 and bicarbonate, occur from algal photosynthesis in most waters with a rich phytoplankton. Owing to the limited rates of CO_2 diffusion across the air-water interface, the deficits persist through night-time more than does oxygen supersaturation.

Levels of pH are chiefly governed by the bicarbonate-carbonate alkalinity and the concentration (or tension) of free CO_2. When the latter is in equilibrium with the atmosphere, reference values of pH are obtained which change only slightly with temperature and increase approximately linearly with the logarithm of the alkalinity. This relationship, and the displacements of pH observed in some Nile waters due to CO_2 enrichment or depletion, are illustrated by Talling & Talling (1966). Thus, the air-equilibrium value is approximately pH 8.0 for Lake Victoria, pH 8.2 for the White Nile near Khartoum, and 8.9 for Lake Albert. The values of pH measured for Nile waters (see Tables 1, 2, and Figs. 90, 93, 94) rarely fall below 7; exceptions include Lake Ambadi and some swamp waters in the Sudd region (Rzóska 1974), and the River Sobat (Talling 1957). In this respect there is a sharp contrast with many other large tropical rivers of lower alkalinity, such as the Amazon and Zaire (Congo). The depression of pH in water of the White Nile (Bahr el Jebel) as it traverses the Sudd region is shown in Fig. 90 (see also Bishai 1962; Kurdin 1968; Rzóska 1974). Conversely, the elevation of pH in regions of the White and Blue Niles with intense algal photosynthesis can be seen in Figs. 85, 92, 93, 94, and 96. Other examples in Lake Nubia-Nasser are noted by Entz in Chapter 19a. The higher values reached are 1 pH unit or more above the air-equilibrium values, as in the White Nile (pH 9.3) and the Blue Nile (pH 9.1–9.2) at Khartoum (Prowse & Talling 1958; Talling & Rzóska 1967; Hammerton 1972d), and Lake George (pH 10+ : Viner 1970). In comparison with these, the ranges of pH recorded by Abdin (1948a) and Ramadan

(1972) at Cairo in the pre-High Dam period are relatively small (Fig. 89).

Plant nutrients

The nutrients considered below are often suspected of limiting algal production in freshwaters, and three can be severely depleted during the growth of phytoplankton in parts of the Nile system. Although the major ions also include important nutrients, which may influence properties of the water as a growth medium, the qualities present are normally much in excess of the amounts incorporated in the algal crops and depletions are not significant. An exception is the requirement for relatively large amounts of carbon dioxide, which often involves appreciable depletions of bicarbonate.

Most analyses of nutrient elements in Nile waters have been confined to the soluble inorganic fraction, from which particulate material was excluded (e.g. by filtration). Some estimations exist of total phosphorus, involving a digestion procedure, mainly for the headwater lakes (Talling & Talling 1965) and a few stations in the Sudan and Egypt (see Table 3). However, as pointed out by Golterman (1975), nutrient transport by rivers will often be dominated by the neglected particulate fraction. Further, the suspended silt of the lower Nile has been shown by Elster & Gorgy (1959) to rapidly influence the concentrations of soluble phosphate and nitrate by adsorption and mineralization.

Phosphate is present in relatively considerable concentrations in much of the Nile, considering its status as a virtually unpolluted river. A major source exists in the Western Rift near Lake Edward. Here a tributary stream rich in particulate phosphorus was found by Golterman (1973), and some others, analysed by Viner (1975), contained > 200 $\mu g/l$ of soluble $PO_4 \cdot P$. The source leads, via the Semliki River (also analysed by Viner 1975), to concentrations around 150 $\mu g/l$ of soluble inorganic, and total, P in Lake Albert (Talling 1963; see Table 3 and Fig. 96). Concentrations downstream are much increased from this supply (Talling 1957a). Some further increase (June) or decrease (December) can occur seasonally in the Sudd swamps, possibly by the control of adsorption equilibria by the level of deoxygenation or by the control of swamp drainage patterns by changing water levels (Talling 1957a; Bishai 1962). At Khartoum, concentrations of soluble ('reactive') orthophosphate were about 100 μg $PO_4 \cdot P/l$ in both the White and Blue Niles before the main period of phytoplankton development (Prowse & Talling 1958; Talling & Rzóska 1967; Hammerton 1972d), but declined thereafter to < 10 μg $PO_4 \cdot P/l$ (Figs. 93, 94). The first part of this decline may be influenced by hydrological changes upstream, such as the continued decline of flood-water in the Blue Nile and the effects of the Sudd and Sobat discharge in the White Nile. This phase included the initial growth of the major diatom *Melosira granulata* in both rivers,

375

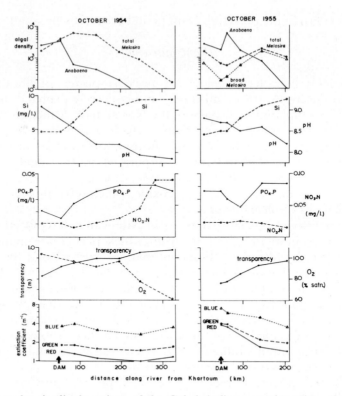

Fig. 92. Two longitudinal sections of the Gebel Aulia reservoir region of the lower White Nile, showing changes in major phytoplankton components and various physical and chemical characteristics influenced by phytoplankton development above the dam. Concentration units for *Melosira granulata* are cells/ml, for *Anabaena flos-aquae* f. *spiroides* coil-turns/ml. (from Prowse & Talling 1958).

which occurred in a relatively phosphate-rich medium. A progressive decline in phosphate and nitrate in the Roseires reservoir was measured by Hammerton (1972b) for the 1967–1968 inter-flood season; these nutrients also decreased along a longitudinal section towards the dam. The later part of the phosphate decline observed by Talling at Khartoum was undoubtedly due to uptake associated with dense phytoplankton, which now included abundant blue-green algae – especially *Anabaena flos-aquae f. spiroides.* Supporting evidence from longitudinal sections upstream on the White Nile is shown in Fig. 92. Besides direct uptake of phosphate by the algae, some adsorption on other suspended particles will occur (Golterman 1975) and may be affected by the higher pH associated with the algal maxima. Equally speculative is a possibly significant exchange of phosphate between water and bottom sediments, especially in such basins of standing water as the Gebel Aulia reservoir and Lake No, for which analyses of sediments show considerable quanti-

376

ties of phosphate-phosphorus (Drover & Bishai 1962; Bishai 1962). Sediment-water exchange in Lake George has been studied by Viner (1969, 1970).

For the Main Nile below Khartoum, there appears to be only fragmentary published information on the spatial or seasonal changes in phosphate concentrations. As the dense phytoplankton passing Khartoum does not persist, its mineralization is likely to be an important source of phosphate. However, it may be eclipsed by the Blue Nile silt, especially in the main deposition zone now in upper Lake Nubia, where blooms of blue-green algae – especially *Microcystis* – seem to be especially dense (cf. Hammerton 1972b; Entz, Chapter 19a). Entz (Chapter 19a) notes some unpublished evidence for a decline in phosphate concentrations from south to north in Lake Nubia-Nasser. Analyses of two 1974 samples from Lake Nasser and Cairo (Table 1) indicate moderate concentrations of total phosphorus. Analyses of reactive inorganic phosphate at Cairo, from November 1941 to August, 1942, are tabulated by Abdin (1948a).

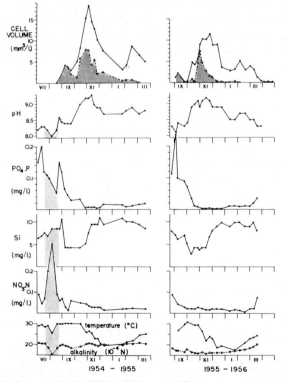

Fig. 93. Seasonal changes in the White Nile near Khartoum, of: phytoplankton crops (with the *Melosira* component shaded), concentrations of some plant nutrients, pH, alkalinity, and temperature. Stippled areas indicate admixture with Blue Nile water. (from Prowse & Talling 1958).

377

The range is 20–100 μg $PO_4 \cdot P/l$, with the lowest values found shortly before arrival of the flood water. Earlier seasonal analyses given by Lucas (1908) were of a similar magnitude, with the highest values from the flood-water period.

Inorganic nitrogen is mainly made up of ammonium- and nitrate-nitrogen; analyses for nitrite have shown insignificant amounts (e.g. Prowse & Talling 1958; Bishai 1962). Concentrations are generally low, and may restrict the growth of some algae unable to fix atmospheric nitrogen (e.g. *Melosira granulata*: see Chapter 27). The methods often used are likely to be inaccurate in the low concentration range (< 100 μg N/l); thus nitrate was probably underestimated by the phenol disulphonic acid method. However, the following general features appear to be well-established.

Nitrate is an exceptional ionic constituent in that its concentration increases steeply with discharge. This feature is connected with its behaviour as a readily leached soil-constituent and probably its sharp pulse of concentration in many African soils after the onset of the rainy season (cf. Viner 1975). Relatively high concentrations, up to 1000 μg $NO_3 \cdot N/l$, have been found in the flood water of the Blue Nile at (Khartoum (Beam 1906, 1908; Talling & Rzóska 1967; see Fig. 94). The River Sobat, which also carries seasonal flood water from Ethiopia,

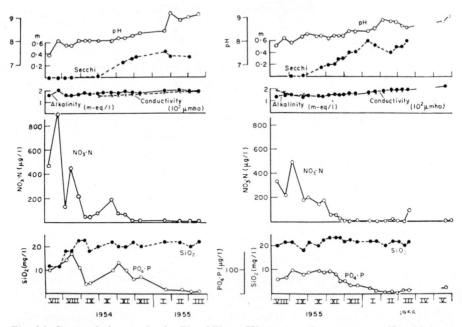

Fig. 94. Seasonal changes in the Blue Nile at Khartoum of transparency (Secchi disc), pH, conductivity (20 °C), alkalinity, and three plant nutrients. (from Talling & Rzóska 1967).

378

has been recorded as rich in nitrate by Kurdin (1968), though not by Talling (1957a) or Bishai (1962). Flood water in Egypt (pre-High Dam period) was also relatively rich in nitrate, as found at Cairo by Abdin (1948a). The decline in concentration at Khartoum after flood levels pass is probably accentuated by the growth of *Melosira granulata*, which in both rivers may be ended by the depletion of nitrate (Prowse & Talling 1958; Talling & Rzóska 1967). A relationship between nitrate depletion and *Melosira* growth is also shown by a longitudinal section along the Gebel Aulia reservoir in October 1954 (Fig. 92). Hammerton (1972b) noted declines in both time and space for the Roseires reservoir in 1967–1968, and Entz (Chapter 19a) refers to evidence for a decline in nitrate concentrations from south to north along Lake Nubia-Nasser.

In the headwaters, excluding the Blue Nile in flood, the measured concentrations of nitrate are generally low ($< 60\,\mu g\,NO_3 \cdot N/l$). Nitrogen-limitation of phytoplankton production may be widespread in the head-water lakes (see Chapter 27) and elsewhere in African freshwaters (cf. Talling & Talling 1965; Moss 1969; Viner 1972). Under stratified lake conditions, higher concentrations of nitrate can appear in the deeper layers of lakes Victoria (Talling 1966) and Albert (Talling 1963; Fig. 96), provided that deoxygenation is not complete.

Ammonium-nitrogen has been generally recorded in concentrations below 100 $\mu g/l$, which are often comparable with the corresponding concentrations of nitrate-nitrogen. A product of organic decay, it is relatively abundant in swamp waters as in the Sudd (Talling 1957a; Bishai 1962; Rzóska 1974), where the main river channel is also enriched. Increases are also typically present in the deeper, less oxygenated layers of stratified lakes, such as Victoria (Talling 1966) and Edward (Beadle 1932; Damas 1938). High concentrations may occasionally appear in surface waters from the decay of dense water-blooms of blue-green algae. One example, which also involved a sequential pulse of nitrate, was observed in fishponds at Gordon's Tree near Khartoum which were filled with water from the nearby White Nile (Talling, unpublished). Metabolic exchanges of $NH_4 \cdot N$ have been studied intensively in L. George, including rapid uptake by the dense phytoplankton (Viner 1972) and diurnal patterns of excretion by zooplankton (Ganf & Blazka 1974).

Silicon is present in dissolved form as silicic acid, $Si(OH)_4$, and (at higher pH) its derivative $SiO(OH)_3^-$. In many analyses it is expressed as the equivalent concentration of silica, the form in which it is in-corporated into diatoms and constitutes the chief biological 'sink' in the Nile system. Concentrations in the headwater lakes, except Lake George, are generally lower than those (typically 10–24 mg SiO_2/l) in the river stretches of the Sudan and Egypt. The difference results partly from depletion by planktonic diatoms in water-bodies of long retention time, and partly from the greater opportunities for dissolution and transfer

379

of ubiquitous rock and soil silicates in running waters. The mobility of silica in tropical soils subject to laterization is well known, and it is perhaps significant that particularly high concentrations of silica are found in the Bahr el Ghazal tributary of the White Nile which drains a lateritic area (Table 3; Talling 1957a). Because of such effects, a close parallelism between the contribution of silicon and phosphate-phosphorus as products of rock weathering – assumed by Golterman (1973, 1975) – seems improbable. A longitudinal section of the White Nile (Talling 1957a) suggested a significant enrichment from the minor torrents which contribute to the main river (Bahr el Gebel) as it descends from the Uganda plateau to the Sudan plains. Enrichment must also occur to the Blue Nile from torrential tributaries in the Gorge region (cf. Table 2).

Depletions induced by diatom growth are conspicuous in several head-water lakes where the prevailing concentrations are relatively low. The annual maximum of *Melosira nyassensis* var. *victoriae* in the offshore waters of Lake Victoria produces an appreciable reduction, from about 4.5 to 3 mg SiO_2/l in 1961 (Talling 1966). Still lower concentrations are frequent in inshore channels and bays where planktonic diatoms (including *M. ambigua*) are abundant, but values below 1 mg SiO_2/l – more likely to limit diatom growth – are not recorded. Such low values do appear typical of Lake Albert (Talling 1963), where diatoms dominate the phytoplankton (see Chapter 27); an alternative abiogenic origin was suggested by Talling but considered improbable by Golterman (1973). In this elongate lake, the concentrations tend to decrease along the axis away from the Semliki inflow (Fig. 96). Downstream, in the Gebel Aulia reservoir, a longitudinal decline was found by Prowse & Talling (1958) during the development of large populations of *Melosira granulata* (Fig. 94). A seasonal depletion occurred in the same month (October) at a station below the dam near Khartoum (Fig. 93), and was of roughly the same magnitude (10 to 5 mg Si/l, or approx. 20 to 10 mg SiO_2/l). No such depletion, attributable to diatom growth, was recognisable at a corresponding station on the Blue Nile during 1953–1956. At Cairo, Abdin (1948a) found that concentrations declined from 12 to 2.4 mg SiO_2/l in the period of 1941–1942 between successive flood-waters, but the part played by diatoms in this is conjectural. Considerably higher concentrations (26 to 13 mg SiO_2/l) are listed by Hurst (1957) for the years 1933–1936. Measurements made after the High Dam do not appear to be published (see, however, Table 1). Iron is present in considerable concentrations in waters of the White Nile in and below the Sudd swamps, where values (approximating total iron in filtered samples) increase sharply in the main river as well as in swamp waters (Talling 1957; Bishai 1962). The increase can be attributed to the release of ferrous iron under reducing conditions and supply of organic complexing substances, the latter probably correlated with an increase in colour.

REFERENCES

Abu Gideiri, Y. B. 1969. The development and distribution of plankton in the northern part of the White Nile. Hydrobiologia 33: 369–378.

Abdin, G. 1948a. Physical and chemical investigations relating to algal growth in the River Nile, Cairo. Bull. Inst. Égypte 29: 19–44.

Abdin, G. 1948b. The conditions of growth and periodicity of the algal flora of the Aswan reservoir (Upper Egypt). Bull. Fac. Sci. Egypt. Univ. No. 27: 157–175.

Abdin, G. 1949. Luminosity measurements in Aswan Reservoir, Egypt. Hydrobiologia 1: 169–182.

Aladjem, R. 1926. Seasonal variation in salinity of Nile water at Rodah (Giza) with special reference to alkaline carbonates. Min. Agr., Tech. & Sci. Serv., Egypt, Bull. 69: 1–11.

Aladjem, R. 1928. Seasonal variation in salinity of Nile water in the Aswan Reservoir and at Rodah (Giza). Min. Agr., Tech. & Sci. Serv., Egypt, Bull. 81: 1–14.

Aleem A. A. & Samaan, A. A. 1969. Productivity of Lake Mariut, Egypt. Part I. Physical and chemical aspects. Int. Revue ges. Hydrobiol. Hydrogr. 54: 313–355.

Bacon, C. R. 1918. Poisoned fish. Sudan Notes & Records 1: 207–209.

Beadle, L. C. 1932. Scientific results of the Cambridge Expedition to the East African Lakes, 1930–31. 4. The waters of some East African Lakes in relation to their fauna and flora. J. Linn. Soc. (Zool.) 38: 157–211.

Beadle, L. C. 1974. The inland waters of tropical Africa. An introduction to tropical limnology. London. Longman. pp. 365.

Beam, W. 1906. Report of the Chemical Section, in Second Report, Wellcome Research Laboratories, Khartoum. Rep. Wellcome trop. Res. Labs. 2: 206–214.

Beam, W. 1908. Nile Waters, in Report of the chemical section of the Wellcome Research Laboratories, Gordon College, Khartoum. Rep. Wellcome trop. Res. Labs. 3: 386–395.

Beam, W. 1911. Sobat River Water, in Fourth Report, Wellcome Tropical Res. Laboratories, Gordon College, Khartoum. Rep. Wellcome trop. Res. Labs. 4: 32–33.

Beauchamp, R. S. A. 1956. The electrical conductivity of the head waters of the White Nile. Nature, Lond. 178: 616–619.

Bini, G. 1940. Ricerche chimiche nelle acque del lago Tana. Missione di studio al lago Tana, vol. III. Part 2, Reale Accademia d'Italia, pp. 9–52.

Bishai, H. M. 1962. The water characteristics of the Nile in the Sudan with a note on the effect of *Eichhornia crassipes* on the hydrobiology of the Nile. *Hydrobiologia* 19: 357–382.

Brook, A. J. & Rzóska, J. 1954. The influence of the Gebel Aulyia dam on the development of Nile plankton. J. anim. Ecol. 23: 101–114.

Carter, G. S. 1955. The papyrus swamps of Uganda. Heffer, Cambridge, England. 25 pp.

Damas, H. 1937. Recherches hydrobiologiques dans les lacs Kivu, Édouard, et Ndalaga. Explor. Parc Albert, Mission H. Damas (1935–1936), fasc. I, 128 pp. Inst. Parcs nat. Congo belge, Bruxelles.

Drover, D. P. & Bishai, H. M. 1962. A preliminary study of some chemical constituents of the bottom deposits of the White Nile. Hydrobiologia 20: 179–184.

Elster, H. J. & Gorgy, S. 1959. Der Nilschlamm als Nährstoffregulator im Nildelta. Naturwissenschaften 46: 147.

Elster, H. J. & Vollenweider, R. A. 1961. Beiträge zur Limnologie Ägyptens. Arch. Hydrobiol. 57: 241–343.

El-Wakeel, S. K. & Wahby, S. D. 1970. Hydrography and chemistry of Lake Manzalah, Egypt. Arch. Hydrobiol. 67: 173–200.

Entz, B. 1972. Comparison of the physical and chemical environments of Volta Lake and Lake Nasser. In Productivity problems of freshwaters (eds. Z. Kajak & A Hillbricht-Ilkowska), Warszawa–Kraków, pp. 883–892.

Fish, G. R. 1957. A seiche movement and its effect on the hydrology of Lake Victoria. Fish. Publ., Lond. No. 10: 1–68.

Ganf, G. G. 1974a. Incident solar irradiance and underwater light penetration as factors controlling the chlorophyll *a* content of a shallow equatorial lake (Lake George, Uganda). J. Ecol. 62: 593–609.

Ganf, G. G. 1974b. Diurnal mixing and the vertical distribution of phytoplankton in a shallow equatorial lake (Lake George, Uganda). J. Ecol. 62: 611–629.

Ganf, G. G. 1975. Photosynthetic production and irradiance-photosynthesis relationships of the phytoplankton from a shallow equatorial lake (Lake George, Uganda). Oecologia 18: 165–183.

Ganf, G. G. & Blczka, P. 1974. Oxygen uptake, ammonia and phosphate excretion by zooplankton of a shallow equatorial lake (Lake George, Uganda). Limnol. Oceanogr. 19: 313–326.

Ganf, G. G. & Horne, A. J. 1975. Diurnal stratification, photosynthesis and nitrogen-fixation in a shallow, equatorial lake (Lake George, Uganda). Freshwat. Biol. 5: 13–39.

Ganf, G. G. & Milburn, T. R. 1971. A conductimetric method for the determination of total inorganic and particulate organic carbon fractions in freshwater. Arch. Hydrobiol. 69: 1–13.

Ganf, G. G. & Viner, A. B. 1973. Ecological stability in a shallow equatorial lake (Lake George, Uganda). Proc. R. Soc. Lond. B. 184: 321–346.

Gay, P. A. 1958. Conductivity of Nile waters. Fifth Ann. Rep. 1957–1958, Hydrobiol. Res. Unit, Univ. Khartoum, pp. 5–7.

Golterman, H. L. 1973. Natural phosphate sources in relation to phosphate budgets: a contribution to the understanding of eutrophication. Water Res. 7: 3–17.

Golterman, H. L. 1975. Chemistry of running waters. In River Ecology, ed. B. Whitton, pp. 39–80. Blackwell, Oxford.

Gorgy, S. 1959. Rate of Evaporation from Lake Quarun. Notes & Mem. Hydrobiol. Dept. Alexandries. 42, 26pp.

Grabham, G. W. & Black, R. P. 1925. Report of the Mission to Lake Tana, 1920–21. Min. of Public Works, Egypt. Government Press, Cairo. 207 pp.

Hammerton, D. 1970. Water characteristics and phytoplankton production. Eleventh Ann. Rep. 1963–1964, Hydrobiol. Res. Unit, Univ. Khartoum, pp. 5–12.

Hammerton, D. 1972a. Survey of work in progress. Blue Nile survey. Fourteenth Ann. Rep. 1966–1967, Hydrobiol. Res. Unit, Univ. Khartoum, pp. 3–11.

Hammerton, D. 1972b. Survey of work in progress. Blue Nile survey. River Nile–Lake Nubia. Fifteenth Ann. Rep. 1967–1968, Hydrobiol. Res. Unit, Univ. Khartoum, pp. 5–14.

Hammerton, D. 1972c. Survey of work in progress. Blue Nile survey. White Nile survey. Sixteenth Ann. Rep. 1968–1969, Hydrobiol. Res. Unit, Univ. Khartoum, pp. 4–9.

Hammerton, D. 1972d. The Nile River – a case history. In: River Ecology and Man, eds. Oglesby, R. T., Carlson, C. A. & McCann, J. A., pp. 171–214. Academic Press, New York & London.

Hecht, A. 1964. On the turbulent diffusion of the water of the Nile floods in the Mediterranean Sea. Bull. Sea Fish. Res. Stn Israel No. 36.

Hurst, H. E. 1925. The Lake Plateau Basin of the Nile. Ministry of Public Works, Egypt. Physical Dept., Paper no. 21, Govt. Press, Cairo. (Section 13: Salt content of the water of the Lake Plateau, pp. 67–73).

Hurst, H. E. 1957. The Nile. 2nd edition. Constable, London.

Hutchinson, G. E. 1957. A treatise on limnology. I. Geography, physics, and chemistry. John Wiley and Sons, New York.

Kurdin, V. P. 1968. Data on hydrological and hydrochemical observations on the White Nile. Inf. Byull. Biol. vnutr. Vod. 2: 49–56. (Russian)

382

Levring, T. & Fish, G. R. 1956. The penetration of light in some tropical East African waters. Oikos 7: 98–109.

Lucas, A. 1908. The chemistry of the River Nile. Survey Dept. Paper no. 7, Govt. Press, Cairo. pp. 1–78.

Morandini, G. 1940. Missione di studio al lago Tana: 3. Ricerche limnologiche, parte prima, Geografia – Fisica. Reale Acad. d'Italia, Rome. 315 pp.

Moss, B. 1969. Limitation of algal growth in some Central African waters. Limnol. Oceanogr. 14: 591–601.

Omar, M. H. & El-Bakry, M. M. 1970. Estimation of evaporation from Lake Nasser. Meteorol. Res. Bull. 2: 1–27.

Omer El Badri Ali. 1972. Sediment transport and deposition in the Blue Nile at Khartoum, flood seasons 1967, 1968 and 1970. M.Sc. Thesis, Univ of Khartoum, unpublished.

Oren, O. H. 1969. Oceanographic and biological influence of the Suez canal, the Nile and the Aswan dam on the Levant Basin. Progr. Oceanogr. 5: 161–167.

Prowse, G. A. & Talling, J. F. 1958. The seasonal growth and succession of plankton algae in the White Nile. Limnol. Oceanogr. 3: 223–238.

Ramadan, F. M. 1972. Characterization of Nile waters prior to the High Dam. Z. Wasser Abwasser Forsch. 5: 21–24.

Rzóska, J. 1957. Conductivity of Nile waters. Fourth Ann. Rep. 1956–1957, Hydrobiol. Res. Unit, Univ. Khartoum, pp. 8–10.

Rzóska, J. 1968. Observations on zooplankton distribution in a tropical river dam-basin (Gebel Aulia, White Nile, Sudan). J. anim. Ecol. 37: 185–198.

Rzóska, J. 1974. The Upper Nile swamps, a tropical wetland study. Freshwat. Biol. 4: 1–30.

Talling, J. F. 1957a. The longitudinal succession of water characteristics in the White Nile. Hydrobiologia 11: 73–89.

Talling, J. F. 1957b. Diurnal changes of stratification and photosynthesis in some tropical African waters. Proc. R. Soc. Lond. B 147: 57–83.

Talling, J. F. 1957c. Some observations on the stratification of Lake Victoria. Limnol. Oceanogr. 2: 213–221.

Talling, J. F. 1963. Origin of stratification in an African Rift lake. Limnol. Oceanogr. 8: 68–78.

Talling, J. F. 1965. The photosynthetic activity of phytoplankton in East African lakes. Int. Rev. ges. Hydrobiol. 50: 1–32.

Talling, J. F. 1966. The annual cycle of stratification and phytoplankton growth in Lake Victoria (East Africa). Int. Rev. ges. Hydrobiol. 51: 545–621.

Talling, J. F. 1969. The incidence of vertical mixing, and some biological and chemical consequences, in tropical African lakes. Verh. int. Verein. Limnol. 17: 998–1012.

Talling, J. F. & Rzóska, J. 1967. The development of plankton in relation to hydro-logical regime in the Blue Nile. J. Ecol. 55: 637–662.

Talling, J. F. & Talling, I. B. 1965. The chemical composition of African lake waters. Int. Revue ges. Hydrobiol. 50: 421–463.

Tottenham, P. M. 1926. The Upper White Nile Mission, 1923. Interim Report. Government Press, Cairo.

Viner, A. B. 1969. The chemistry of the water of Lake George, Uganda. Verh. internat. Verein. Limnol. 17: 289–296.

Viner, A. B. 1970. Ecological chemistry of a tropical African lake. Ph.D. thesis, Univ. London.

Viner, A. B. 1972. Responses of a mixed phytoplankton population to nutrient enrichments of ammonia and phosphate, and some associated ecological implications. Proc. R. Soc. Lond. B 183: 351–370.

Viner, A. B. 1975. The supply of minerals to tropical rivers and lakes (Uganda). In: An introduction to land-water interactions (ed. Olsen, J.) Chapter 10. Springer-Verlag. pp. 227–261.

Viner, A. B. & Smith, I. R. 1974. Geographical, historical and physical aspects of Lake George. Proc. R. Soc. Lond. B. 184: 235–270.

Vollenweider, R. A. 1960. Beiträge zur Kenntnis optischer Eigenschaften der Gewässer und Primärproduktion. Mem. Ist. Ital. Idrobiol. 12: 201–224.

Worthington, E. B. 1930. Observations on the temperature, hydrogen-ion concentration and other physical conditions of the Victoria and Albert Nyanzas. Int. Rev. ges. Hydrobiol. Hydrogr. 24: 328–357.

27. PHYTOPLANKTON: COMPOSITION, DEVELOPMENT AND PRODUCTIVITY

by

J. F. TALLING

Introduction

Several reasons may lead one to expect vigorous and varied developments of planktonic algae in regions of the Nile system. First, there are large headwater lakes in which lacustrine phytoplankton can develop, and possibly travel down their outflows as potential 'inocula' for renewed growth downstream. Second, the retention of water in the reservoirs of Sudan and Egypt provides – at least seasonally – the additional time favourable to phytoplankton development. Third, the great length of the river and its component stretches increases the time of travel of any water-mass, and so the opportunities for planktonic growth.

In this section an outline is given of general features of the phytoplankton found in various regions of the Nile system. Although some description of the species-composition of these communities is provided, the main emphasis is upon the patterns of development in space and time. Where available, estimates of production rates are also discussed. Many regional details can also be found in other chapters. The impact of phytoplankton development on chemical characteristics of the river water is briefly surveyed in chapter 26.

Regional development

THE HEADWATER LAKES

a. *Lake Victoria*

The phytoplankton of Lake Victoria (Fig. 95) is probably more varied, and richer in species, than that of any other part of the Nile system. This diversity was recognised early in the hydrobiological exploration of the Nile system; it is illustrated in the systematic accounts and lists of species published by Schmidle (1899, 1902), West (1907), Ostenfeld (1908, 1909), Virieaux (1913), Woloszyúska (1914), Hustedt (1922), Bachmann (1933), Thomasson (1955), Talling (1966) and Richardson (1968). Most of these descriptions were based on preserved samples obtained as isolated collections by plankton nets, liable to selective over-representation of some components and the loss of some fragile or very small

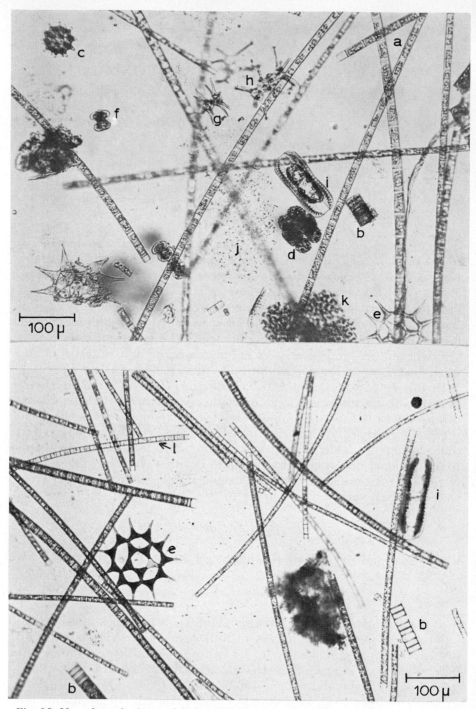

Fig. 95. Net phytoplankton of Lake Victoria from A) northern offshore water, B) Pilkington Bay near Jinja. The algae include: a. *Melosira nyassensis* var. *victoriae*; b. *Melosira agassizii*; c. *Coelastrum cambricum*; d. *Coelastrum reticulatum*; e. *Pediastrum clathratum*; f. *Staurastrum muticum*, g, h. *Staurastrum limneticum*; i. *Surirella nyassae*; j. *Aphanocapsa* sp. (?); k. *Microcystis* sp.; l. *Melosira ambigua*. (from Talling 1966).

species. Quantitative estimations over prolonged periods were initiated by Fish (1957), and extended by Talling (1957c, 1966) and Evans (1962). They were based on northern areas of the lake near the Nile outflow at Jinja, and showed large differences – qualitative and quantitative – between the more offshore 'open' lake and the inshore waters of the numerous gulfs and bays. However, Talling did not find such significant differences between the various offshore regions.

Diatoms and blue-green algae make up the greater part of the phytoplankton in both inshore and offshore waters. Green algae are also well represented offshore by numerous species of Chlorococcales (esp. *Pediastrum clathratum, Coelastrum reticulatum, C. cambricum, Sorastrum americanum, Tetraedron arthrodesmiforme*) and desmids (esp. *Staurastrum leptocladum* f. *africanum, S. limneticum, S. anatinum, S. gracile* var. *nyansae, S. muticum, Cosmarium moniliforme*), although their contribution to the total biomass is usually minor. Species of the diatom genus *Melosira* are often dominant in both inshore and offshore regions. The cosmopolitan *M. ambigua* is usually typical of inshore bays, but *M. nyassensis* var. *victoriae* is the major form in the main lake, where it shows large seasonal changes of population density related to the annual cycle of stratification (Fish 1957; Talling 1966).

Offshore, low densities during the warmest and most stratified phase are also shared by many other species in the phytoplankton, and are reflected in the seasonal minimum of chlorophyll *a* content. This condition is ended by the cooler and windy phase of June to August, when near isothermal mixing returns to most of the lake, filaments of *Melosira* are returned to the surface layers, and deep accumulations of nutrients are dispersed. In 1961, population increases were then exhibited by most planktonic algae, and most conspicuously by the principal diatoms (*Melosira nyassensis* var. *victoriae, M. agassizii, Nitzschia acicularis, Surirella nyassae, Stephanodiscus astraea*). A quite different type of population cycle was followed by the blue-green algae *Anabaena flos-aquae* and *Anabaenopsis tanganyikae*, which declined over the isothermal period and increased strongly during the following phase of superficial stratification. Over the year as a whole, however, the predominant blue-green algae were very small-celled colonial forms, tentatively identified as *Aphanocapsa elachista* and *A. delicatissima* West & West (Talling 1966).

The total concentration of phytoplankton in the offshore surface waters is not large. Expressed as chlorophyll *a* content, Talling (1966) obtained values of 1.2–5.5 mg chl-*a*/m³. Much larger concentrations occur in shallow inshore bays and gulfs; the large Kavirondo Gulf is particularly rich, with 20 mg chl-*a*/m³ recorded in December 1960 (Talling 1965).

b. *Lake Kioga*

Traversed by the Victoria Nile, this lake is extremely shallow, dendritic

387

West (1913) and Bachman (1933), but were not found by Talling or other later observers.

A few scattered records were made by Prowse and Talling (unpublished) on the content of phytoplankton in the inflowing and outflowing White Nile. Very little was found in the inflow, below the Murchison Falls, but the two characteristic diatoms of L. Albert were present along a considerable stretch of the outflowing White (Albert) Nile.

d. *Lake Tana*

Very little is known about the phytoplankton of this large, relatively shallow lake on the headwaters of the Blue Nile. Brunelli & Cannicci (1940) give a brief list of species and photomicrographs; some additional, mostly unpublished observations were made later by Talling (cf. Talling & Rzóska 1967, p. 644). Seasonal and quantitative records are lacking, except for a single estimation by Talling of chlorophyll *a* content (3.7 mg/m^3) from near Bahar Dar in March 1964.

Brunelli & Cannicci found a scanty development of blue-green algae, chiefly of the genera *Anabaena* and *Microcystis*, but more considerable numbers of diatoms, especially species of *Melosira* and *Surirella*. In Talling's samples, collected in March 1964 from near the south shore, a diatom resembling *Melosira granulata* var. *jonensis* f. *procera* was strongly dominant, although a spirally coiled *Anabaena* sp. was common. Both he and the Italian hydrobiologists found that some desmids (esp. *Staurastrum leptocladum*), *Pediastrum clathratum*, and *Surirella* spp. were well represented, as in some other large African lakes including L. Victoria. Detached littoral diatoms were also common, at least in inshore areas, as might be expected in this shallow and often turbulent lake.

The Sudan plain

Observations on both the White and Blue Niles indicate that very little of the headwater lacustrine phytoplankton survives and prospers after the descent to the Sudan plain. Qualitatively, most of the dominant species of lakes Victoria, Albert, and Tana do not reappear conspicuously in the lower stretches. Quantitatively, the concentrations of cells counted in samples from the points of entry to the Sudan plain near Juba (White Nile) and Roseires (Blue Nile) were very low. In 1964, the Tana plankton was scantily represented at the Tissisat Falls, and not seen at all at a station further down the Blue Nile Gorge.

After entering the plain, the White Nile flows over shallow gradients for about 1400 km before it is seasonally impounded in the Gebel Aulia reservoir near Khartoum. In this reach, scattered observations have shown phytoplankton to be present in low densities, near or below the limits of quantitative estimation. It is composed predominantly of the

tions are possible but conjectural. Silicon limitation is very unlikely because of the relatively high residual concentrations, even though the predominant diatom of the Nile – *Melosira granulata* – is generally associated with considerable concentrations (Kilham 1971).

From these and other African inland waters, there is evidence that strongly alkaline conditions (near or above pH 9.0) are unfavourable for *Melosira* – plankton (Talling & Talling 1965; Talling & Rzóska 1967). Such conditions are reached as the result of photosynthesis during the seasonal maxima of blue-green algae in the White and Blue Niles near Khartoum. It is possible, but not proven, that they may limit the *Melosira* component, although vigorous photosynthesis by the blue-green algae continues (Prowse & Talling 1958).

Some attempts to identify limiting nutrients in Lake Victoria by bio-assaytype experiments are described by Fish (1956) and Evans (1961).

Rates of photosynthetic production

As noted in Chapter 26, the photosynthetic activity of phytoplankton may alter the content of dissolved oxygen, carbon dioxide, and hence pH. These changes have a strong diurnal component, related to daily irradiation, which was traced as early as 1927 in the Kavirondo Gulf of Lake Victoria (Worthington 1930). Following observations by Pyle (1950) on standing waters in the Sudd region, Talling (1957b) extended them for a lagoon and other productive water-bodies of the White Nile system, and derived estimates of the daily gross photosynthetic production per unit area. The latter lay between 4 and 11 g $O_2/m^2 \cdot$ day, approximately equivalent to between 1.5 and 4 g $C/m^2 \cdot$ day, and were compared with independent estimates from water samples exposed in light and dark bottles.

The light and dark bottle (oxygen) method was used by Talling (1957, 1966) to examine the conditions of production in a number of headwater lakes, including Victoria, Albert, Edward, and George. Later and more detailed work on L. George is described by Ganf (1975) and Ganf & Horne (1975). In all these lakes estimates of maximum daily gross production were high, between 10–16 g $O_2/m^2 \cdot$ day; net photosynthesis was difficult to estimate, but was probably much lower (cf. Ganf 1974 for L. George). Talling (1966) showed that the photosynthetic rates per unit area were relatively insensitive to the wide variation in phytoplankton density (per unit volume), mainly because of self-shading effects on light penetration. However, the maximum (light-saturated) rates of photosynthesis per unit water volume were closely correlated with population density. The connecting factor, the maximum specific activity per unit population, was notably high compared to general experience with phytoplankton.

Further down the White Nile, in and below the Gebel Aulia reservoir near Khartoum, intense photosynthetic activity during phytoplankton maxima was studied by Talling (1957b), Prowse & Talling (1958), and Hammerton (1972a). Some of the resulting depth-profiles of activity, with related factors, are shown in Figs. 100 and 101. The photosynthetic zone is typically compressed into a layer 1 to 2 m deep, partly due to self-shading behaviour but mainly to the fine suspended material characteristic of the White Nile. As in the headwater lakes, the maximum specific activity (per unit of population) was usually very high; the rates calculated per unit cell volume agreed broadly with later measurements

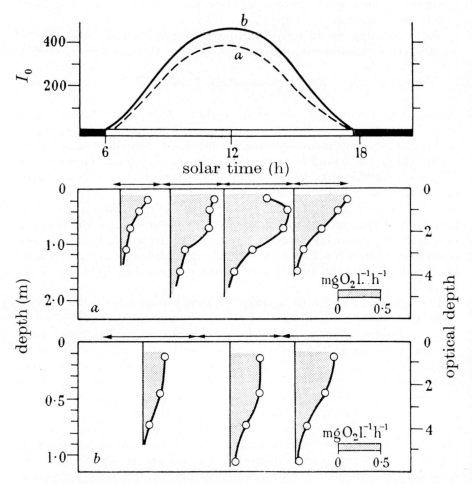

Fig. 100. Depth profiles of photosynthetic rate (stippled areas) in the Gebel Aulia reservoir for various periods during the day, during (a) 12 December 1954 (b) 7 October 1955. The diurnal variation of incident solar irradiance I_0 (in kerg/cm$^2 \cdot$ s = W/m^2) is shown above. (from Talling 1957b).

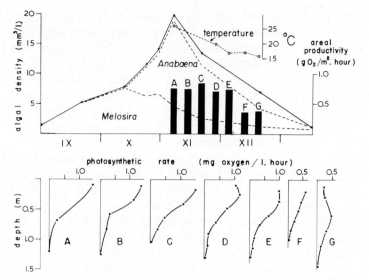

Fig. 101. Growth and decline of the 1953–1954 phytoplankton community in the White Nile at Gordon's Tree near Khartoum. Densities (in mm³ cell volume/l) of two major components are shown, and associated rates of photosynthetic production per unit area (histograms). The latter are calculated from depth profiles of photosynthetic rate shown below (from Prowse & Talling 1958).

from the lakes based upon chlorophyll a content. The gross production rates per unit area were moderately high, often exceeding 0.5 g $O_2/m^2 \cdot h$ or an estimated 5 g $O_2/m^2 \cdot$ day. For example, over a period of 30 days during the 1953 population maximum illustrated in Fig. 101, the estimated areal rates were 0.77 ± 0.07 g $O_2/m^2 \cdot h$ or 2.4 ± 0.2 g $C/m^2 \cdot$ day. Later work in 1965, by D. Hammerton and his collaborators, yielded estimates of similar magnitude (Hammerton 1972a).

Hammerton (1972a, b, c) has also measured photosynthetic production at various points, and in various seasons, on the Blue Nile. The results are discussed and illustrated (Figs. 60, 61) in Chapter 16. The maximum rates per unit area are even higher than in the White Nile, possibly due to the reduced silt content and greater light transmission.

Few measurements have been published for regions of the Nile below Khartoum. Elster & Vollenweider (1961) list a series of 7 stations (^{14}C exposures) from the (old) Aswan reservoir to Cairo, with the maximum areal rate (1.06 g $C/m^2 \cdot 6h$) and deepest photosynthetic zone (5 m) in the Aswan reservoir. Corresponding values at Cairo were 0.358 g $C/m^2 \cdot 6h$ and 2 m. More intensive work was carried out by Vollenweider (1960; Elster & Vollenweider 1961) and Aleem & Samaan (1969) on Lake Mariut, an extremely productive and polluted delta lake near Alexandria (see Chapter 20). Here the dense phytoplankton could

399

reduce the photosynthetic zone to as little as 0.35 m, but very high estimates of daily production (> 4 g C/m² · day) were obtained.

Under present-day conditions on the Main Nile proper, the highest areal rates of photosynthetic production are to be expected from Lake Nubia-Nasser, where dense phytoplankton may occur with low background turbidity. Few measurements are published, but Entz (1972) refers to one high estimate of 15.5 g O_2/m² · day (or ~ 5.8 g C/m² · day); similarly high, as yet unpublished estimations by Samaan (3.2–5.2 g C/m² · day) are noted in Chapter 19.

In a more general setting, the production rates measured in the Nile system are notable for the high values per unit area often reached, and high specific rates per unit of population seem widespread. Although the depth of the photosynthetic zone varies over a wide range, as from about 15 m in Lake Victoria to 1 m in the White Nile near Khartoum, self-shading effects appear less influential than the often high 'background' light absorption. High specific rates of population growth are indicated by some seasonal observations, especially during the first phase of annual water storage in reservoirs on the Blue and White Niles. Further evidence can be obtained from spatial increases downstream (Talling & Rzóska 1967), which involve the interrelationship between events in time and space – probably the most fundamental issue of river biology.

REFERENCES

Abdin, G. 1948a. Physical and chemical investigations relating to algal growth in the River Nile, Cairo. Bull. Inst. Égypte 29: 19–44.

Abdin, G. 1948b. Seasonal distribution of phytoplankton and sessile algae in the River Nile, Cairo. Bull. Inst. Égypte 29: 369–382.

Abdin, G. 1948c. The conditions of growth and periodicity of the algal flora of the Aswan reservoir (Upper Egypt). Bull. Fac. Sci. Egypt. Univ. no. 27: 157–175.

Abdin, G. 1949. Biological productivity of reservoirs, with special reference to Aswan Reservoir (Egypt). Hydrobiologia 1: 469–475.

Aleem, A. A. & Samaan, A. A. 1969. Productivity of Lake Mariut, Egypt. Part II. Primary production. Int. Rev. ges. Hydrobiol. Hydrogr. 54: 491–527.

Bachmann, H. 1933. Phytoplankton von Victoria Nyanza-, Albert Nyanza- und Kiogasee. Ber. Schweiz. bot. Ges. 42: 705–717.

Bachmann, H. 1936. The fishery grounds near Alexandria. XIV Phytoplankton from the Nile. Fisheries Res. Directorate, Cairo No. 22, 2pp.

Beadle, L. C. 1974. The inland waters of tropical Africa. An introduction to tropical limnology. London. Longman. pp. 365.

Bini, G. 1940. Ricerche chimiche nelle acque del lago Tana. Missione di studio al lago Tana, vol III. Part 2, Reale Accademia d'Italia, pp. 9–52.

Brook, A. J. 1954. A systematic account of the phytoplankton of the Blue and White Nile. Ann. Mag. nat. Hist., Ser. 12, 7: 648–656.

Brook, A. J. & Rzóska, J. 1954. The influence of the Gebel Aulyia dam on the development of Nile plankton. J. Anim. Ecol. 23: 101–114.

Brunelli, G. & Cannicci, F. 1940. Le caratteristiche biologiche del Lago Tana. Missione di Studio al Lago Tana. Volume Terzo. Ricerche limnologiche. Parta Seconda. Chimica e biologia. Roma: Reale Accademia d'Italia. pp. 71–132.

Elster, H. J. & Vollenweider, R. A. 1961. Beiträge zur Limnologie Ägyptens. Arch. Hydrobiol. 57: 241–343.

Entz, B. 1972. Comparison of the physical and chemical environments of Volta Lake and Lake Nasser. In Productivity problems of freshwaters (eds. Z. Kajak & A. Hillbricht-Ilkowska), Warzawa–Kraków, pp. 883–892.

Evans, J. H. 1961. Growth of Lake Victoria phytoplankton in enriched cultures. Nature, Lond. 189: 417.

Evans, J. H. 1962. The distribution of phytoplankton in some Central East African waters. Hydrobiologia 19: 299–315.

Fish, G. R. 1956. Chemical factors limiting growth of phytoplankton in Lake Victoria. E. Afr. agric. J. 21: 152–158.

Fish, G. R. 1957. A seiche movement and its effect on the hydrology of Lake Victoria. Fish. Publ., Lond. No. 10: 1–68.

Ganf, G. G. 1974. Rates of oxygen uptake by the planktonic community of a shallow equatorial lake (Lake George, Uganda). Oecologia 15: 17–32.

Ganf, G. G. 1975. Photosynthetic production and irradiance-photosynthesis relationships of the phytoplankton from a shallow equatorial lake (Lake George, Uganda). Oecologia, 18: 165–183.

Ganf, G. G. & Horne, A. J. 1975: Diurnal stratification, photosynthesis and nitrogen-fixation in a shallow, equatorial lake (Lake George, Uganda). Freshwat. Biol. 5: 13–39.

Grönblad, R. 1962. Sudanese Desmids II. Acta Bot. Fenn. 63: 3–19.

Grönblad, R. Prowse, G. A. & Scott, A. M. 1958. Sudanese Desmids. Acta Bot. Fenn. 58, 82 pp.

Hammerton, D. 1970. Water characteristics and phytoplankton production. Eleventh Ann. Rep. 1963–1964, Hydrobiol. Res. Unit, Univ. of Khartoum, pp. 5–12.

Hammerton, D. 1972a. Studies of primary production in the river Nile. Thirteenth Ann. Rep., 1965–1966, Hydrobiol. Res. Unit, Univ. of Khartoum, pp. 16–17.

Hammerton, D. 1972b. Blue Nile survey; River Nile–Lake Nubia. Fifteenth Ann. Rep., 1967–1968, Hydrobiol. Res. Unit, Univ. of Khartoum, pp. 5–14.

Hammerton, D. 1972c. The Nile River – a case history. In River Ecology and Man, pp. 171–214. Academic Press, New York & London.

Hustedt, F. 1922. Bacillariales. In: Schröder, B. Zellpflanzen Ostafrikas. Hedwigia 63: 117–173.

Kilham, P. 1971. A hypothesis concerning silica and the freshwater planktonic diatoms. Limnol. Oceanog. 16: 10–18.

Ostenfeld, C. H. 1908. Phytoplankton aus dem Victoria Nyanza. Sammelausbeute von A. Bogert, 1904–1905. Bot. Jb. 41: 330–350.

Ostenfeld, C. H. 1909. Notes on the phytoplankton of Victoria Nyanza, East Africa. Bull. Mus. comp. Zool. Harv. 52: 171–181.

Prowse, G. A. 1954. Phytoplankton. First Ann. Rep. 1953–1954, Hydrobiol. Res. Unit, Univ. College of Khartoum, pp. 12–14.

Prowse, G. A. & Talling, J. F. 1958. The seasonal growth and succession of plankton algae in the White Nile. Limnol. Oceanogr. 3: 223–238.

Pyle, J. 1950. In: Problems of Fisheries, (ed.) H. Sandon. Sudan Notes and Records 32: pp. 21–22.

Richardson, J. L. 1968. Diatoms and lake typology in East and Central Africa. Int. Rev. ges. Hydrobiol. 53, 299–338.

Rzóska, J. 1958. Notes on the biology of the Nile north of Khartoum. Fifth Ann. Rep., 1957–1958, Hydrobiol. Res. Unit, Univ. of Khartoum, pp. 16–20.

Rzóska, J. 1974. The Upper Nile swamps, a tropical wetland study. Freshwat. Biol. 4: 1–30.

Rzóska, J., Brook, A. J. & Prowse, G. A. (1955). Seasonal plankton development in the White and Blue Nile near Khartoum. Verh. int. Ver. theor. angew. Limnol. 12: 327–334.

Salah, M. & Tamas, G. 1970. General preliminary contribution to the plankton of Egypt. Bull. Inst. Oceanogr. Fish. Alexandria 1: 305–337.

Schmidle, W. 1899. Die von Professor Dr. Volkens and Dr. Stuhlmann in Ost-Afrika gesammelten Desmidiaceen. Bot. Jb. 26: 1–59.

Schmidle, W. 1902. Das Chloro- und Cyanophyceenplankton des Nyassa und einiger anderer inner-afrikanischer Seen. Bot. Jb. 33: 1–33.

Talling, J. F. 1957a. The longitudinal succession of water characteristics in the White Nile. Hydrobiologia 11: 73–89.

Talling, J. F. 1957b. Diurnal changes of stratification and photosynthesis in some tropical African waters. Proc. R. Soc. Lond. B 147: 57–83.

Talling, J. F. 1957c. Some observations on the stratification of Lake Victoria. Limnol. Oceanogr. 2: 213–221.

Talling, J. F. 1963. Origin of stratification in an African Rift lake. Limnol. Oceanogr. 8: 68–78.

Talling, J. F. 1965. The photosynthetic activity of phytoplankton in East African lakes. Int. Rev. ges. Hydrobiol. 50: 1–32.

Talling, J. F. 1966. The annual cycle of stratification and phytoplankton growth in Lake Victoria (East Africa). Int. Rev. ges. Hydrobiol. 51: 545–621.

Talling, J. F. & Rzóska, J. 1967. The development of plankton in relation to hydrological regime in the Blue Nile. J. Ecol. 55: 637–662.

Thomasson, K. 1955. A plankton sample from Lake Victoria. Svensk bot. Tidskr. 49: 259–274.

Virieux, J. 1913. Plancton du lac Victoria Nyanza. In: Voyage de Ch. Alluaud and R. Jeannel en Afrique orientale (1911–1912). Résultats scientifiques. Paris. 20 pp.

Vollenweider, R. A. 1960. Beiträge zur Kenntnis optischer Eigenschaften der Gewässer und Primärproduktion. Mem. Ist. Ital. Idrobiol. 12: 201–244.

Wawrik, F. 1959. Beitrag zur Planktonkunde Ober-Aegyptens. Anz. Ost. Akad. Wiss. (1959) 15: 300–306.

West, G. S. 1907. Report on the freshwater algae, including phytoplankton, of the third Tanganyika Expedition conducted by Dr. W. A. Cunnington, 1904–1905. J. Linn. Soc. (Bot.) 38: 81–197.

West, G. S. 1909. Phytoplankton from the Albert Nyanza. J. Bot., Lond. 47: 244–246.

Woloszynska, J. 1914. Studien über das Phytoplankton des Viktoriasees. Hedwigia 55: 184–223.

Worthington, E. B. 1929a. Report on the fishery survey of lakes Albert and Kioga. London, Crown Agents for the Colonies. 136 pp.

Worthington, E. B. 1929b. The life of Lake Albert and Lake Kioga. Geogr. J. 74: 109–132.

Worthington, E. B. 1930. Observations on the temperature, hydrogen-ion concentration and other physical conditions of the Victoria and Albert Nyanzas. Int. Rev. ges. Hydrobiol. Hydrogr. 24: 328–357.

EPILOGUE

by

J. RZÓSKA

In this monographic book we have tried to explain the extraordinary character of the Nile through its history of three main phases: the geological past, – the ecological changes in its environment affecting man and life generally, – and finally the phase of exploration, scientific enquiry and human possession. Some of the authors, including the editor, have written their contributions from a distance of time and space; this deletes details but brings out the important general features.

The title of our monograph describes the Nile as 'ancient'. It is so in its origin, dating back to the Miocene and probably older. In a different time scale the Nile valley in the north harbours a sequence of fluviatile civilisations, spaning at least 30,000 years and unsurpassed in continuity and intensity. In stark contrast, the East African region of the river's headwaters does not show such concentration of human activity, recorded by history. No climatic and other environmental pressure, as in the north, existed here. But although 'history' is thin, prehistory is richer than elsewhere.

The Nile is also 'young' in the linkage of river sectors of different age and origin. Although dates are still disputed, the recent, present, river system is said to be not older than 30,000 years (ch. 1–3). During that time climates along the river have changed and have moved gradually into a phase of stark aridity over 3,000 kilometers of its northern stretch. Though shifts of rainbelts created at times savanna conditions even in the Sahara, with eloquent testimonies left by hunting and pastoral people (ch. 4a), yet by the fourth millenium BC even Egypt began to feel the advent of desert conditions, the savanna disappeared and man had to move near to permanent water, the river. It is soon afterwards that attempts of water management started, for example on the water courses in the eastern desert, then still temporarily active; now only traces of dried up wadis and khors are marked on maps.

The desert conditions, lasting now for several thousands of years, have interrupted the biological allegiance of Egypt with the rest of Africa, though flora and fauna bear a varying testimony to former links (ch. 6). Significantly, it is the aquatic life which forms the strongest bonds as exemplified by phytoplankton, zooplankton and fishes.

The river pierces the desert and like a life artery brings in the water supply from distant Ethiopia and East Africa. The hydrographic map of the Nile basin is wondrous to contemplate (Fig. 14). It looks like a slender, enormous plant with its 'roots' inside Africa and its 'crown' in

Egypt, thousands of kilometers to the north. With their own supply either negligible or strictly seasonal, the Sudan and Egypt live literally on foreign water.

The whole river system traverses half of Africa. Its south-north orientation and wide span of latitudes (0–32) is unique amongst major rivers of the tropics and subtropics. In its course the Nile flows through 3 climatic zones, a number of different landscapes and very different geomorphological formations. Each of these environments exerts some influences on the river; most spectacular are the Upper Nile swamps and the Blue Nile Gorge. The Gorge region of the Blue Nile has provided the sediment which has built up the alluvial Nile valley and the Delta. New and startling work on the 'age and rates of denudation' of the Gorge (McDougall, Morton & Williams, 1975, Nature 254: 207–209) has revealed the age of the four top lava flows, giving the lower limit for the start of incisions of both the Blue Nile and the Atbara at 27–23 millions of years B.P. Calculations made on the volume of sediments deposited along the river valley with the bulk in the Delta vary but are probably in the region of $150,000 \pm 40,000$ km³. The rate of erosion in the uplands of Ethiopia, which has supplied 98% of the sediment, has accelerated recently and deforestation is one of the cause. The authors stress that Ethiopian sediments and pollen are noticeable in the Delta 'during at least part of the Cainozoic'.

Chapters 1 and 2 of our book have suggested such previous connections between Ethiopia and the Egyptian Nile; the new work confirms these suggestions. The space photographs used in our book show this long process of sediment deposition most eloquently.

I owe Dr. M. Williams much gratitude for drawing my attention to the above work; he read also critically some chapters dealing with palaeoecology. Much what has been written here on this subject will be more competently treated in the forthcoming publication on the 'Prehistoric Occupation in the Central Sudan' ed. by J. D. Clark & M. Williams. Another book which has just been published is 'Fish Communities in Tropical Freshwaters, their distribution, ecology and evolution' by R. H. Lowe-McConnell, Longman, London 1975. This contains a great amount of information also on the Nile system, comparing it with other tropical waters.

To return to our book, it is important to remember that the geographical exploration and establishment of the map of the Nile was only accomplished in broad details about 1875, one hundred years ago. Some of the numerous reports written by many explorers have enabled us to reconstruct to some degree the state of the river before human interference. The phase of scientific enquiry started with water supply problems for Egypt; the hydrological survey of the Nile basin is unsurpassed anywhere in the world and it all started because of dire necessity.

But this is a book on the biology of the river; all other aspects form the background to the real purpose of this book, but they are necessary.

Hydrobiological exploration of the Nile system started in the second half of the 19th century both in Egypt and the East African Lake Plateau. In east Africa the earliest seem to have been the Germans in their newly acquired colonies; the first plankton net was dipped into Lake Victoria in 1888. Many other expeditions followed, notably that from Cambridge University in 1930 under E. B. Worthington, all doing pioneer work. This book does not attempt to record the history of hydrobiological work in Africa, though it would be of great interest. In the Sudan and Ethiopia such work was initiated later; in the Sudan it was concentrated at the Gordon College, later the University of Khartoum. In Ethiopia such work started with the Italian efforts in the late thirties.

But concentrated and sustained scientific work can only be done by permanent institutions, where experience accumulates steadily. Such institutions exist now along the river: on lake Victoria we have the Jinja station in Uganda, founded in 1946/47 as part of the East African Fisheries Research Organisation; in the Sudan the Hydrobiological Research Unit of the University of Khartoum, founded in 1953, is active, in Egypt there are research centres at Aswan and at Alexandria; all of these are in close touch with government fishery agencies. Their work will have to be more closely coordinated in view of the new tasks imposed by the new river regimes caused by dams.

Our books gathers the existing results of work on the biology of the Nile and mainly its fundamental features. We are reasonably well informed about the chemical and physical characteristics of the river system. The phytoplankton and the primary production has been investigated in a number of points. Animal life, zooplankton and fishes are known in main outlines; not so benthos with its implications for medical and social problems. A great number of problems remain unsolved, for example the *Eichhornia* menace, the repercussions of the newly created dam basins at Roseires and above all the Aswan High Dam basin, which are in the first stages of biological maturation. They will influence the extent of insect 'pets' and the spread of mollusc vectors of schistosomiasis. Sediments now falling out in Lake Nubia, instead of Egypt, will have to be studied because they will create new environments in the Sudan. These are some few of the tasks for the future. To these must be added the further development of fisheries; a special chapter was contemplated for our book but this was abandoned in view of more competent attempts by FAO and other agencies, with resources far beyond our possibilities. We have used fishery data as indicators of biological productivity for example in the chapters on Lake Nasser and Lake Nubia. But in hydrobiology applied research and practical measures are closely interlinked with fundamental recognition. Although much of the previous work recorded here has been overtaken by the advent of

405

the complete management of the river, yet any further and expected changes will have to be checked against previous conditions.

Above all, the Nile even in its present form lives in bouts of high and lower intensities in its sectors. Its life forms a world of its own and nowhere is the contrast stronger than in the long desert sector. There a rich life exists in the waters with a multitude of delicate organisms while the surrounding landscape is largely lifeless.

Finally, some remarks on the controversial issue of reservoirs. Thousands have been built in the last decades for economic and sometimes political reasons. Their ecological effects have been discussed in numerous symposa and in a spate of articles and books. Few of the great dam basins have elicited more heated arguments than the Aswan High Dam. We must remember in this respect some important points about Egypt, where the dam has been built. Only 3% of the area of almost one million km² of Egypt is fertile soil, the rest is desert. Human density is 600 per km² in the cultivated land and only one per 7 km² outside. These figures impose grave decisions on any government.

The side effects of this and others dams often adverse have been recognised and debated. We will not enter into these controversies, more competent and powerful agencies are involved. The Nile dams are a fact and biologists will have to cope with the problems arising. Cooperation of the adjacent countries along the Nile is necessary and working groups will have to be established to deal effectively with the tasks ahead.

This book might be helpful.

AUTHORS INDEX

SUBJECT INDEX

The detailed Chapter Contents give information which is not repeated here. The components of the complex Nile system are treated as separate items.

paper manufacture, 192
Pistia fauna, 213
Plankton, see river components
  in rivers, conditions for development, 385, 396
  influence of dams, 215, 217, 244
phytoplankton, see particular river components
  general survey, 385ff
  blooms, 377
  effect on transparency, 361, 363
  survey of sectors, see Chapter Contents
  L. Kioga, 387–388
  L. Victoria, 385–387
Photosynthetis production
  L. Victoria, Albert Edward George, 397
  Gebel Aulia, 398
  Khartoum, White and Blue Nile, 398, 399
  Aswan Dam (previous) 399; L. Mariut, 399
  Nasser lake, 400

Qarun, Birket el, 47, 48, 360, see also Faiyum

Radiocarbon dates, 5, 8, 14, 16, 21, 22, 26, 35, 37, 38, 39, 42, 43, 47, 48
Rahad river, 12, 262
Rainfall, 34, 37, 43, 79, 80 (map)
Rivers see particular items
  biological development of plankton, main basis see Time/Space
  levels ancient, 5–7
Rockdrawings, 38ff, 43
Roseires dam basin, 245ff
  benthic molluscs, 255
  impact on phytoplankton, 251–253, 361
    on zooplankton, 254–255
  light penetration, 250, 361, 363
  oxygen conditions, 248, 249
  map and illustrations p. 219, 246, 247, 248
  sediments, 158
  thermal stratification, 248, 361, 363
  water characteristics, 248–250, 358, 363, 373, 376
Rudolf lake, 25, 26, 27, 28, 127, 128, 131, 137, 230

Sabaloka, (VI cataract), 15, 16, 263
Sahara, 45, 108, 113, 130
Salt marshes, Egypt, 51
Sediments, 155ff
  ancient, 4, 5, 7

Blue Nile contribution, 13, 157, 158
  composition, 4, 7, 8, 157, 159
  cores, 22ff
  fall out, 17, 89, 156, 157, 159, 160, 279
  fertility, 158, 159, 310, 311
Schistosomiasis
  ancient record, 321
  distribution, 321ff
  threat for future, 323
  vectors, 322, 323
Sennar dam, 149, 244, 252, 255, 358
Shaheinab, 16, 42
Shabona, 42
Singa, 42
Sobat river, 14, 21, 215, 374, 375
Swamps
  East Africa, distribution and character, 177, 178
    altitudinal, 184
    effect on traffic, 182
    fauna, 190
    influence on water characteristics, 186ff, 189
    management and use, 190, 191, 194
    papyrus, 179ff
    productivity, 192ff
    vegetation, 177, 182ff
    zonation, 181ff
  Upper Nile ('Sudd'), 14, 27, 43, 75, 77, 202
    fauna, 106, 211, 213, 215
    fishes and fisheries, 211, 371
    flooding, 202ff, 205, 207
    human ecology, 203, 207ff
    micro relief, 206
    plankton, 208, 209, 210
    vegetation, 203, 204, 205
    water characteristics, 208, 209, 365, 374
    zooplankton, 209, 210

Tana lake, 223ff
  bathymetric map, 224
  climate, 225
  discharge, 227
  drainage basin, 223
  evaporation, 236, 360
  fishes, 127, 128, 131, 132
  hydrology, 226
  limnology, 226ff
  mollusca, 112, 229, 230
  origin, 220, 404
  plankton, 229, 390
  pollen analysis, 26
  space photo, 225

416